Signals and Communication Technology

More information about this series at http://www.springer.com/series/4748

Xinpeng Xing · Peng Zhu
Georges Gielen

Design of Power-Efficient Highly Digital Analog-to-Digital Converters for Next-Generation Wireless Communication Systems

Springer

Xinpeng Xing
Graduate School at Shenzhen
Tsinghua University
Shenzhen
China

Georges Gielen
Departement Elektrotechniek,
ESAT-MICAS
Katholieke Universiteit Leuven
Leuven
Belgium

Peng Zhu
Zhongguancun Dongsheng Technology Park
Analog Devices, Inc.
Beijing
China

ISSN 1860-4862 ISSN 1860-4870 (electronic)
Signals and Communication Technology
ISBN 978-3-319-66564-1 ISBN 978-3-319-66565-8 (eBook)
https://doi.org/10.1007/978-3-319-66565-8

Library of Congress Control Number: 2017950276

Printed on acid-free paper

This Springer imprint is published by Springer Nature
The registered company is Springer International Publishing AG
The registered company address is: Gewerbestrasse 11, 6330 Cham, Switzerland

To my wife Qin!

Preface

Mobile devices motivate the continuous development of the communication industry, meaning low-power designs are in great demand. Analog-to-digital converters (ADCs) build bridge between analog front-end and digital cores, and play a more important role in emerging transceiver architectures. On the other hand, due to CMOS technology scaling, the ADC design suffers from design issues such as decreasing headroom voltage. Recently, one popular direction of ADC design is shifting more functions from the voltage and the analog domains to the time and the digital domains, by using a voltage-controlled oscillator (VCO). The implementation of circuits in the time domain immediately takes advantage again of the technology scaling with reduced gate delay.

This book is a result of our research work in ESAT-MICAS, K.U. Leuven, in the field of highly digital ADC design. Recently, highly digital ADC has drawn more and more interest from both academy and industry, and the increasing number of publication on this topic is an evidence. We believe this is a natural evolution when IC designers are facing ongoing-scaling CMOS technologies. With this common interest, however, we cannot find systematic book on this topic but only papers and book chapters as far as we know. It has been our motivation to present our work in highly digital ADC as a book. This collection is mainly based on our own five chip designs in ESAT-MICAS, and also, some previous contributions by various researchers to this field are also included.

This book focuses on the systematic design of power-efficient highly digital ADC for future communication applications, with both architecture- and circuit-level innovations. The first two designs are 40MHz-BW 12bit CT $\Delta\Sigma$ ADCs implemented in 90nm CMOS, with quantizations done by highly digital VCO-based quantizer. Various circuit-level techniques are applied for power reduction, including the shaped switched capacitor digital-to-analog converter (SC DAC), the look-up-table (LUT)-based digital calibration, and the current-sharing feedforward-compensated OTA. The third design is a 40MHz-BW two-step open-loop VCO-based ADC in 40nm CMOS. With hardware-economic structure and mostly digital building blocks, an excellent FoM of 42fJ/step is obtained. However, only first-order noise shaping is realized for the whole ADC, limiting its

SNR performance. To go further, in our fourth design, a nonlinearity cancellation 0–2 MASH ΔΣ ADC structure with innovative dual-input VCO-based quantizer is adopted. The optimized systematic parameters and the highly digital circuit blocks extend the FoM of the state-of-the-art high-bandwidth ΔΣ ADCs to 35fJ/step. With second-order noise shaping, the ADC performance is improved, and one analog front-end integrator is still needed, however. In the final design of this work, a VCO-based integrator is proposed to replace the power-hungry analog integrator in the traditional ΔΣ ADC topology, realizing second-order ΔΣ ADC without any analog integrator. 74-dB SFDR and a FoM of 52fJ/step over a 40MHz bandwidth have been achieved in the demonstration.

With the state-of-the-art power efficiencies, the presented highly digital ADCs are very suitable for the applications of next-generation wireless communication standards, including but not limited to 802.11n and LTE. Furthermore, the design methodology and design innovations described in this book could also be applied to ADCs for other applications.

Shenzhen, China
Beijing, China
Leuven, Belgium
March 2017

Xinpeng Xing
Peng Zhu
Georges Gielen

Acknowledgements

At the moment to archive our work, we could not forget the people, who contributed directly or indirectly to this book. First, we would like to thank editor Oliver Jackson, Springer UK, for his careful and patient organization work during the whole publication procedure. We also would like to express our gratitude to anonymous reviewers for their valuable comments and suggestions for improvement. During our PhDs in K.U. Leuven, Prof. Jean Berlamont, Prof. Michiel Steyaert, Prof. Marc Moonen, Prof. Wim Dehaene, Prof. Pieter Rombouts, Prof. Patrick Wollants, Prof. Guy Vandenbosch, all in K.U. Leuven, and Prof. Michael Peter Kennedy, University College Cork, and Andrea Baschirotto, University of Milano-Bicocca served kindly as the program committee members, many thanks to them for their valuable feedback and communications on our work. We would like to thank Danielle Vermetten, Ben Geeraerts, Chris Mertens, Frederik Daenen, Noella Gaethofs, and Michel De Cooman in MICAS-ESAT for their outstanding secretariat and logistic works. Without their supports, we cannot accomplish the whole chip design. We also would like to thank our ex-colleagues (also friends) Yi Ke and Peng Gao for the technical discussion in data converters and so on.

Besides, we would like to convey our thanks to the Funds for Scientific Research Flanders (FWO-V.), Belgium, and the European Union Seventh Framework Programme for Research and Technology Development (EU-FP7) for the project financial supports.

Finally, our most grateful thanks are for our families, for their invaluable love, support, and patience. Also many thanks to our colleagues and friends, for the wonderful time and memory in Leuven.

Shenzhen, China
March 2017

Xinpeng Xang

Contents

Abbreviations

2G	Second Generation
3G	Third Generation
AAF	Anti-Aliasing Filtering
AC	Alternating Current
A/D	Analog-to-Digital
ADC	Analog-to-Digital Converter
ADSL	Asymmetric Digital Subscriber Line
AFE	Analog Front-End
AGC	Automatic Gain Control
APWM	Asynchronous Pulse-Width Modulator
ASIC	Application-Specific Integrated Circuit
BB	BaseBand
BPF	Band-Pass Filter
BW	Bandwidth
CF	Crest Factor
CM	Common Mode
CMFB	Common-Mode Feedback
CMOS	Complementary Metal Oxide Semiconductor
CS	Current-Steering or Current-Sharing
CT	Continuous-Time
D/A	Digital-to-Analog
DAC	Digital-to-Analog Converter
DC	Direct Current
DCM	Duty Cycle Modulation
DEM	Dynamic Element Matching
DFF	Direct Feedforward
DNCF	Digital Noise Cancellation Filter
DNL	Differential Nonlinearity
DR	Dynamic Range
DSM	Delta-Sigma Modulator

DSP	Digital Signal Processing
DT	Discrete-Time
DVD	Digital Versatile Disc
EDA	Electronic Design Automation
ELD	Excess Loop Delay
ENOB	Effective Number of Bit
ERBW	Effective Resolution Bandwidth
FET	Field Effect Transistor
FF	FeedForward or Flip-Flop
FIR	Finite Impulse Response
FM	Fading Margin
FoM	Figure of Merit
FS	Full Scale
GBW	Gain bandwidth product
GP	General Purpose
GSM	Global System for Mobile Communications
HSDPA	High Speed Downlink Packet Access
IC	Integrated Circuits
ICO	Current-Controlled Oscillator
IF	Intermediate Frequency
IIT	Impulse Invariant Transformation
IM2/3	2nd/3rd-order Intermodulation Distortion
INL	Integrated Nonlinearity
IPTV	Internet Protocol Television
I/Q	In-phase/Quadrature
ISI	Inter-Symbol Interference
ISSCC	International Solid-State Circuits Conference
LFSR	Linear Feedback Shift Register
LHP	Left Half Plane
LMS	Least Mean Square
LNA	Low-Noise Amplifier
LO	Local Oscillator
LPF	Low-Pass Filter
LSB	Least Significant Bit
LTE	Long Term Evolution
LUT	Look-Up Table
LVDS	Low-Voltage Differential Signalling
MASH	Multi-stAge noise SHaping
MDAC	Multiplying DAC
MIM	Metal-Insulator-Metal
MOS	Metal-Oxide-Semiconductor
MSB	Most Significant Bit
NAND	Negated AND

NF	Noise Figure
NFC	Near Field Communication
NM	Noise Margin
NMOS	N-channel MOSTFET
NRZ	Non Return-to-Zero
NTF	Noise Transfer Function
OSR	Oversampling Ratio
OTA	Operational Transconductance Amplifier
PA	Power Amplifier
PC	Personal Computer
PCB	Printed Circuit Board
PD	Phase Detector
PLL	Phase-Locked Loop
PMOS	P-channel MOSFET
PM	Phase Margin
PSD	Power Spectral Density
PSRR	Power Supply Rejection Ratio
PVT	Pcocess Voltage Temperature
PWM	Pulse-Width Modulator
RF	Radio Frequency
RMS	Root-Mean-Square
ROM	Read-Only Memory
RSR	Receivable Signal Range
RZ	Return-to-Zero
SA	Sense-Amplifier
SAFF	Sense-Amplifier Flip-Flop
SAR	Successive Approximation Register
SAW	Surface Acoustic Wave
SC	Switched-Capacitor
SFDR	Spurious-Free Dynamic Range
S/H	Sample and Hold
SiGe	Silicon-Germanium
SNDR	Signal-to-Noise-Distortion Ratio
SNR	Signal-to-Noise Ratio
SoC	System on Chip
SP	Standard Performance
SQNR	Signal-to-Quantization-Noise Ratio
STF	Signal Transfer Function
TDMA	Time Division Multiple Access
THD	Total Harmonic Distortion
TSPC FF	True-Single-Phase Clock Flip-Flop
TV	Television
VCO	Voltage-Controlled Oscillator

VDSL	Very-High-Bit-Rate Digital Subscriber Line
V-F	Voltage-Frequency
VGA	Variable Gain Amplifier
VGLNA	Variable Gain Low-Noise Amplifier
VTC	Voltage-to-Time Converter
WLAN	Wireless Local Area Network
XOR	Exclusive OR

List of Figures

List of Tables

Chapter 1
Introduction

1.1 Background and Motivation

Integrated circuits (IC) is a multi-disciplinary and multi-level industry, which is composed of product definition, system design, circuit design, electronic design automation (EDA), chip fabrication, package, and test. As a connection between system design and chip fabrication, circuit design needs to accommodate the needs of both ends. For communications, both communication standards, the system architecture and the widely-used complementary metal-oxide-field-semiconductor (CMOS) technology have huge impacts on the circuit design.

1.1.1 Communication Evolution

1.1.1.1 High Date Rate and High Bandwidth

When we look back on the communication history, two facts are very evident and accompany the whole communication evolution: higher data rates and more mobile devices. Being the most important driving sources for developments in the communication and IC industries for the past decades, it is safe to say that these trends will continue in the future years.

For both wireline and wireless communications, a higher data rate is the primary characteristic to define a new generation communication standard. With a higher data rate, communications can be improved in two aspects. Firstly, more information can be transferred at the same time, so some new services are becoming reality. For instance, video phone can be supported in 3G communications, while only audio and messaging are available with 2G GSM systems. Secondly, for data with the same size, less waiting time is needed for the transfer. For example, to download a 4.7 GB

X. Xing et al., *Design of Power-Efficient Highly Digital Analog-to-Digital Converters for Next-Generation Wireless Communication Systems*, Signals and Communication Technology, https://doi.org/10.1007/978-3-319-66565-8_1

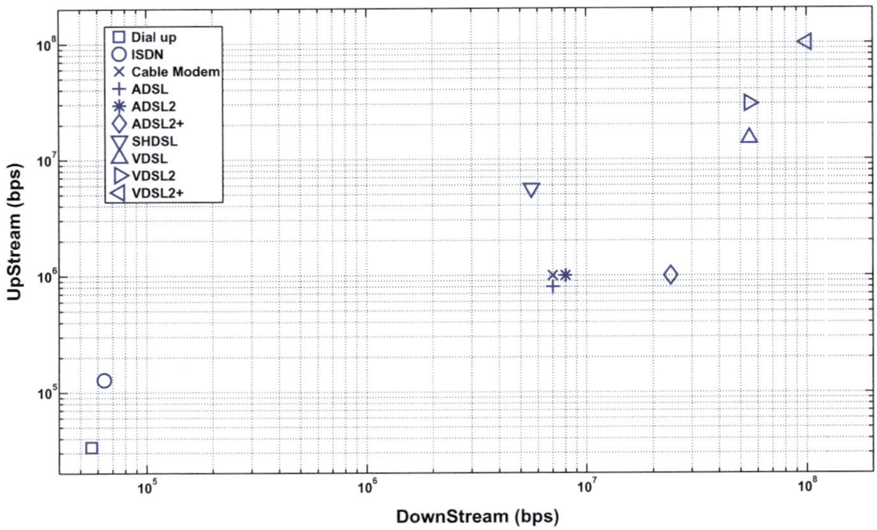

Fig. 1.1 Data rate comparison between different wireline internet accesses

DVD from internet, it takes about 1.5 h with ADSL access versus only 6 min with a VDSL2+ connection.

In Fig. 1.1, the typical downstream and upstream data rates of different wireline internet accesses are plotted. The data rate is increased by about 2000× from the old Dial-up (1990s) to the latest VDSL2+ (2006). Furthermore, the bit rate of optical communication can reach 10 Gbps [1]. Similar stories happen with wireless communications, as shown in Fig. 1.2. For cellular systems, at least 5 generations of standards have been defined, and the data rate has been upgraded by more than 10000×.

Finally, it is also a trend to combine several previously separated communication services together. For example, ATT U-verse provides IPTV (several channels), IP telephone and broadband internet services with a single cable [2]. This combination simplifies the connection but calls for a larger data rate for the single connection. With the same modulation, higher communication data rates require higher analog bandwidths, setting new challenges for analog and mixed-signal circuit designs. It is obvious that analog-to-digital converters (ADCs) with tens of MHz bandwidths are in great demand to realize the next generation of communication systems, including LTE, WLAN 802.11n and VDSL2.

1.1.1.2 Low Power for Mobile Devices

The ultimate goal of communication developments is that anybody can be connected anytime and at anyplace. To achieve this objective, mobile devices are indispensable for the whole communication system. Wireless communication not only expands its

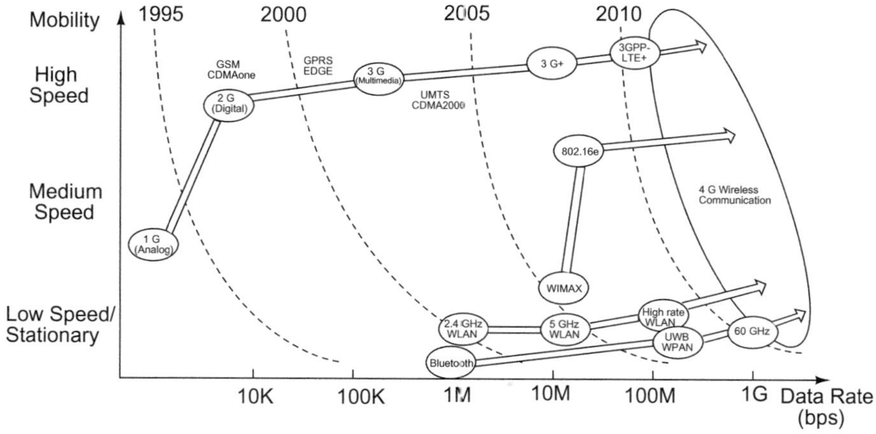

Fig. 1.2 Evolution of wireless communications [3]

data rate faster than cable communication, but it also costs less time to be a widely accepted mature technique. Previously, with the invention of the GSM, more and more telephones shifted from wireline to wireless communications. Nowadays with the birth of the smartphone and tablet, more and more tasks are done with these mobile devices, including email, internet browning and TV. In Fig. 1.3, the relative global shipments of four consumer electronics (smartphones, tablets, desktop and laptop PCs) are plotted from 2012 to 2017 (estimated values for 2013 and 2017) [4]. It is shown that the smartphone and tablet shipments have exceeded by far those of PCs. Considering the large gap between the smartphone/tablet and cellphone (about 1.4 billion in 2010)/laptop sales, the growth in shipments of smartphones and tablets will continue in future years.

Compared to cellphones and laptops, there are many more mobile connections possible with smartphones and tablets, including LTE, WLAN, NFC and HSDPA. While more connections and more functions integrated, also a longer operation time is expected for smartphones and tablets. With limited battery volume, low power is therefore a more stringent design criterion for chips used in smartphones and tablets. In recent highly digital receiver design, the ADC is the performance bottleneck and consumes a large part of the total power. Based on the above description and deduction, low-power ADCs with tens of MHz bandwidths are key blocks for next-generation smartphone and tablet developments.

1.1.2 CMOS Technology

Since its invention in 1967 [5], CMOS technology has been used to implement most of the ICs, especially Systems on Chip (SoC). The primary reason for this

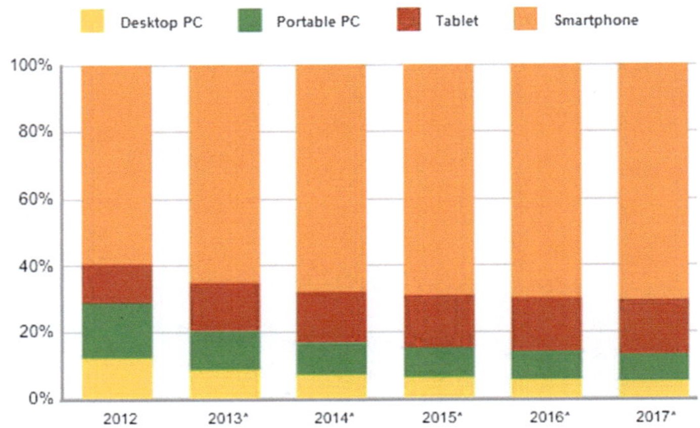

Fig. 1.3 Relative global shipments of four consumer electronics from 2012 to 2017 (estimated values for 2013 and 2017) [4]

fact is the low cost of the CMOS technology. In 2003, one could buy 10 million CMOS transistors with only $1. Another cause of the wide application of CMOS is the constant technology scaling, as shown in Fig. 1.4. In 1965, Moore Gordon predicted the exponential growth of the number of transistors on one IC [6]. With technology scaling, the minimum gate length of the CMOS transistor is reduced by 30% and the transistor area is halved every 18 months. As a result, the chip area, the parasitic capacitance and the power consumption reduce correspondingly. Although the investment cost of the scaled process line increases, the price of a function block still decreases continuously due to this scaling. At the same time, with reduced parasitic capacitance and increased oxide capacitance density, the CMOS transistor can work at a higher frequency, increasing the circuit bandwidth and throughput.

While bringing lots of benefits for the IC industry, CMOS technology scaling also introduces lots of design challenges into the chip design. For digital circuits, the noise margin is becoming smaller and the circuit is more sensitive to noise and interferences, increasing the circuit bit error rate. Furthermore, with thinner gate oxide, the leakage current has increased, which has exceeded the dynamic power in some advanced technologies.

In analog and mixed-signal circuit designs, the first design difficulty comes from the voltage. To avoid transistor breakdown, the CMOS power supply voltage decreases continuously. However, in order to prevent severe leakage problems and to accommodate the doping and temperature variations, the transistor threshold voltage scaling is much slower, as shown in Fig. 1.5. This means that the available signal headroom voltage is becoming smaller and smaller. If the same SNDR is needed for an analog circuit, more power consumption is needed to reduce the noise and distortion levels. Another potential problem induced by the voltage scaling is the switch. With a smaller supply voltage, the switch on-resistance becomes larger and its linearity becomes worse [8]. This introduces another challenge for example switched-capacitor

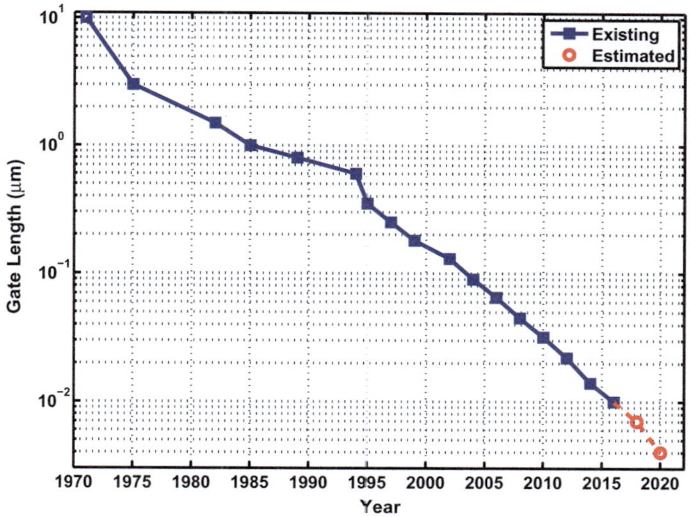

Fig. 1.4 CMOS technology scaling over the last 45 years and prediction for the next 4 years

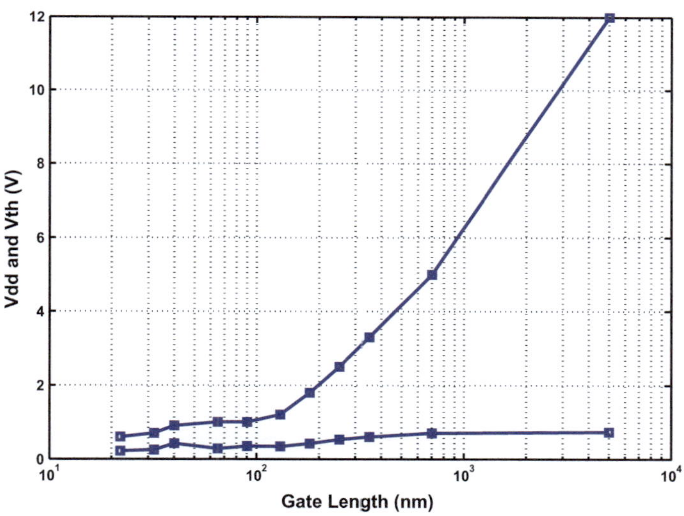

Fig. 1.5 Decreasing supply voltage and relatively constant threshold voltage with CMOS scaling [7]

circuit design. Although in some technologies more than one threshold voltage is provided for different kinds of circuits, additional mask layer are required in such fabrication, leading to a higher cost.

Secondly, the output impedance of a CMOS transistor in saturation is proportional to its gate length. If the minimum gate length is adopted to make use of the technology scaling, then an amplifier's intrinsic gain is decreasing with gate length scaling:

$$A_{DC} = g_m r_o = \frac{2I_d}{V_{gs} - V_{th}} \cdot \frac{V_E L}{I_d} = \frac{2V_E L}{V_{gs} - V_{th}} \tag{1.1}$$

For example, in 65 nm CMOS, if the minimum gate length used, the intrinsic gain of a single-stage amplifier is only 16.5 dB, and it can not exceed 28dB even when a ×16 minimum gate length is used [9]. Moreover, with smaller headroom voltages, some techniques are excluded in circuit design to improve the amplifier gain, e.g. cascode structures and gain boosting.

Passive components also suffer from the technology scaling. The overall stacking and individual metal heights are decreasing, so both the parasitic resistance and the vertical parasitic capacitance increase. These result in worse capacitor and inductor quality factors [10]. All the mentioned challenges should be overcome when designing low-power ADCs with tens of MHz bandwidths in advanced CMOS technologies, by introducing innovations at both architecture and circuit level.

1.1.3 Wireless Receiver Architectures

In wireless communications, besides the functionality, also the power consumption, the integrity (bulk) and the programmability are important design criteria when evaluating a transceiver. To achieve such goals, different transceiver architectures have been proposed, including the heterodyne receiver [11], the homodyne receiver (direct conversion receiver) [12], the low-IF receiver [13], the direct sampling receiver [14] and the subsampling receiver [15]. The general trends for these receiver architectures are that they use less off-chip RF components and realize more functions on chip. For the on-chip circuits, less RF and analog blocks but more functions are implemented digitally. The reasons underneath this fact are that on-chip and digital solutions show their advantages in price, robustness and bulk. As a result, more design challenges are shifted to the data converters, and ADCs with better specifications are required to connect the receiver analog front-end (AFE) and the digital part.

In the heterodyne receiver, there are at least two mixers and the design of each block (including the ADC) is relaxed. However, it is a complicated system and part of filters (for image suppression) can not be implemented in CMOS technology. The alternative off-chip solution leads to problems of huge bulk, parasitics, 50Ω impedance matching and cost, making it not suitable for future mobile communications.

If the sampling and A/D conversion are moved from the baseband (BB) stage to the intermediate frequency (IF) stage, a heterodyne receiver evolves into a digital IF receiver. In a digital IF receiver, the AFE is simplified and the mismatch between the IF I/Q paths is mitigated. However, the ADC needs to work at IF and not BB, and the choice of the IF frequency is dictated by the ADC's maximum speed. In the digital IF receiver, since less filtering is performed preceding the ADC, a larger ADC dynamic range (DR) is required to avoid overloading.

In the homodyne receiver, since the local oscillator (LO) frequency is the same as the RF frequency, there is no image problem anymore, leading to a very simple RF stage. Also only a low-performance ADC is required in the homodyne receiver. However, the homodyne structure suffers from several nonidealities and generally it is not suitable for high-performance receivers. Because of the parasitic coupling between the two inputs of the mixer, a DC-offset is generated at the mixer output, saturating the following stages. In some systems, the DC-offset is constant and can be removed by digital calibration. However, in some systems (e.g. TDMA systems), the DC-offset is different for different channels and bursts, which makes digital calibration difficult. Another similar design challenge in the homodyne receiver is the flicker noise of the active components, especially for some communication standards with small bandwidth.

The low-IF receiver combines the advantages of both the heterodyne and the homodyne receivers: full integration and good immunity to the DC-offset and flicker noise. Because of these advantages, the low-IF receiver can reach a better performance than the homodyne receiver. Since the selectivity is executed by the digital filter in the low-IF receiver, an ADC with high DR is always required. Furthermore, if a higher-accuracy ADC is applied, then the preceding polyphase bandpass filter can be replaced by a simple 1st-order lowpass filter, resulting into a digital low-IF receiver.

One design trend of modern receivers is to put the ADC closer to the antenna, making the AFE simpler. If the sampling and A/D conversion are placed at the RF stage, the direct sampling receiver [14] is developed, which has the simplest AFE. In this architecture, after some amplification and filtering, the RF signal is sampled and converted into a digital stream directly. There is no analog mixer and most functions are shifted to the digital domain. In most direct sampling receivers, only a small partition of the whole bandwidth is occupied by the desired signal, so a bandpass ADC is preferred to save power consumption. The mostly digital characteristic makes the direct sampling receiver very suitable for advanced CMOS technology scaling and an excellent choice for high-reconfigurability receivers. However, the ADC and the preceding variable gain amplifier (VGA) operating at RF are difficult to realizes. Moreover, with little signal conditioning circuits, the DR requirement of the ADC is much higher compared to those in other receivers. These two requirements make the ADC design in existing CMOS technologies a huge challenge. Finally, with the high input signal frequency and DR requirement, the jitter performance of the clock generation circuit is very stringent. These ADC and PLL design challenges make the direct sampling receiver almost impossible to realize for communication standards with high performance in existing CMOS technologies.

In the subsampling receiver, the ADC speed requirement is relaxed compared to the direct sampling receiver. However, the requirements of the AFE and the sample-and-hold (S/H) remain the same. Moreover, if the input frequency is m× higher than the sampling frequency, 2 m× more noise is folded into the Nyquist band during sampling. At the output of the S/H, the phase noise is increased by $m^2\times$, which makes this architecture very sensitive to clock jitter.

1.2 The Research Objective of the Book

The ADC is an indispensable block in most electronic systems, including but not limited to consumer products, telecommunications, instrumentation, automotive and military applications. Since digital signal processing (DSP) is becoming more economic and more powerful, the AFE is becoming simpler and the ADC is playing a more important role. At the same time, the ADC design needs to face the challenges from new product definition standards and the CMOS technology scaling.

As one important high-accuracy ADC structure, the delta-sigma modulator (DSM) ADC (i.e. oversampling ADC) is widely used in many applications such as audio and sensing. In applications with 10–100 MHz bandwidths and 10–12 bit accuracy, previously these A/D conversions were mostly done by pipeline ADCs. With their good figure of merit (FoM) and intrinsic anti-aliasing filtering, the continuous-time (CT) DSM ADC is considered potentially being a substitute of the pipeline ADC. Thus it is meaningful to investigate the design challenges of low-power CT DSM ADCs and enable them to be applicable in the next-generation products. From 1997 to 2011, the CMOS technology scaled down by 18× (from 0.8 μm to 45 nm), however the DSM ADC bandwidth increased by 2604× (from 48 kHz to 125 MHz), and the corresponding FoM is improved by 43×. In the future, the technology scaling is believed not to be so fast as before and maybe 10 nm is the end. It is clear that innovations at both architecture level and circuit level are required for the further bandwidth enhancement and power reduction of DSM ADCs.

This book is about innovative highly-digital ADC design, targeting at high-bandwidth (30–50 MHz) moderate-resolution (10–12 bit) ADCs for next-generation communications, with as low as possible power consumption. For low-power design, one popular solution is to keep the architecture simple and to avoid high-performance analog blocks. To maintain the desired resolution, different necessary digital calibrations are added. To be compatible with deep nano CMOS technology and meet the low-power requirement of mobile electronic systems, a popular trend of ADC design is to shift more and more functions from the analog domain and the voltage domain to the digital domain and the time domain, by using voltage-controlled oscillators (VCO) and pulse-width modulators (PWM) in the ADC. By using VCO-based time-domain quantization, it is possible to reach 1st-order noise shaping with an open-loop structure, solving the stability issue of traditional high-order DSMs thoroughly. For the DSM ADCs based on VCOs, nonlinearity of the VCO voltage-frequency transfer function is the ADC performance bottleneck and should be solved primarily.

In this book, the design and silicon implementation of 5 oversampling ADCs are discussed. Highly-digital low-power ADCs for next-generation wireless communications will be proven. The major contributions are summarized as below, and will be presented with more detail from Chaps. 5 to 8.

1. A two-step open-loop VCO-based ADC architecture is proposed. This architecture realizes 1st-order noise shaping with only three performance-relaxed and mostly digital blocks. The VCO nonlinearity is mitigated by distortion cancellation and input swing reduction techniques. The DAC matching requirement is also relaxed by the implicit dynamic element matching (DEM) of the frequency-type VCO-based quantizer. The silicon implementation of the ADC is presented. The excellent figure-of-merit (FoM) of the demonstrator in 40 nm CMOS technology is 42fJ/Step.
2. A 2nd-order 0–2 multi-stage shaping (MASH) DSM ADC with fully-VCO-based integrator and quantizers in 40 nm CMOS technology is proposed and the design details are discussed, realizing a high-order noise shaping DSM ADC without any analog amplifier. Measurement results with a state-of-the-art FoM (52fJ/Step) are presented and discussed.
3. The systematic design of a 2nd-order 0–2 MASH DSM ADC over a 40-MHz bandwidth in 40 nm CMOS technology is presented. A dual-input VCO-based quantizer topology is proposed to implement a low-power multi-input adder and integrator, with no penalty in terms of nonlinearity. The silicon implementation of the ADC is presented. The measured result extends the FoM of the state-of-art high-bandwidth DSM ADCs to 35 fJ/Step.
4. A current-shaped switched-capacitor (SC) DAC, a capacitive local feedback and a current-sharing feedforward (FF) OTA are proposed to reduce analog power consumption in closed-loop CT DSM ADCs. The systematic design of two 4th-order 40 MHz-bandwidth 12-bit DSMs in 90 nm CMOS technology is presented, measurement results are also given.

1.3 The Book Organization

This book mainly covers the architecture design and circuit optimization of highly-digital low-power DSM ADCs for next generation wireless communications. The book is organized as follows:

Chapter 2 gives an introduction of popular ADC architectures, including flash, two-step, pipelined, successive approximation register (SAR) and DSM ADCs. The basic operation principle of each architecture is briefly presented; the performance and speed limitations are also given. A brief comparison between different ADC architectures is provided. Finally the application of ADCs in communication systems is discussed. The relationship between the ADC accuracy specification and the receiver performance requirements is also given in detail.

Chapter 3 talks about CT DSM ADCs. First the linear analysis of a DSM is given in short. Then different structural options for DSMs are discussed and compared, including single-bit and multi-bit quantizations, 1st-order and higher-order noise shapings, feedback and feedforward loop filter topologies, single-loop and MASH structures, MASH-0 and 0-MASH topologies. The last part of this chapter focuses on building block nonidealities of CT DSM ADC. The nonideality modelling and corresponding solutions are also presented.

Chapter 4 provides an overview of existing low-power VCO-based ADCs, including closed-loop DSM ADCs with VCO-based quantizers and open-loop oversampling ADCs. First operation principles and characteristics of VCO-based quantizations are given. Since the VCO nonlinearity is the bottleneck of highly-digital ADC performance, the main content of this chapter is about how to solve the VCO nonlinearity issue in VCO-based ADC design.

Chapter 5 describes the behavioral and circuit designs of two 40MHz-BW 12-bit CT DSM ADCs with VCO-based quantizer in 90 nm CMOS technology. Also their measurement results are presented. The first design uses a shaped SC DAC to relax the OTA slew-rate requirement. A LUT-based calibration is also applied to eliminate the multi-bit DAC distortion. Both system-level optimization (half-period delay) and circuit-level improvements (capacitive local feedback and current-sharing FF OTA) are applied in the second design to achieve lower power.

Chapter 6 proposes an open-loop two-step VCO-based ADC architecture. It realizes 1st-order noise shaping with only three highly digital blocks, mitigating the VCO nonlinearities and relaxing the DAC matching requirement. A 40 MHz-BW 10-bit ADC implemented in 40 CMOS achieves 59.5 dB SNDR and shows an excellent FoM of 42fJ/Step.

To address the VCO nonlinearity and circuit power challenges, design innovations are implemented at architecture and circuit levels in Chap. 7. A nonlinearity-cancellation technique in a 0–2 MASH VCO-based DSM ADC is proposed, a dual-input VCO-based quantizer topology is also proposed to implement a low-power multi-input adder and integrator, with no penalty in terms of nonlinearity. Fabricated in a 40 nm CMOS process, a proof-of-concept prototype achieves 66.8-dB SNDR with a 40 MHz bandwidth, consuming only 4.98 mW. This result extends the FoM of state-of-the-art high-bandwidth DSM ADCs to 35 fJ/Step.

To realize a mostly-digital DSM ADC structure with high-order noise shaping, a VCO is proposed to replace the power-hungry analog integrator in the traditional DSM ADC topology. In Chap. 8, we report a 0–2 MASH DSM ADC with fully-VCO-based integrator and quantizers, without any analog integrators. Taking advantage of the highly-digital structure and relieving the VCO nonlinearity bottleneck, a 74-dB SFDR and a FoM of 52 fJ/Step over a 40 MHz bandwidth are achieved.

Chapter 9 summarizes and concludes the research work on power-efficient highly-digital ADC design. Some suggestions for future work are also proposed.

References

1. I.A. Young, E. Mohammed, J.T.S. Liao, A.M. Kern, S. Palermo, B.A. Block, M.R. Reshotko, P.L.D. Chang, Optical I/O technology for tera-scale computing solid-state circuits. IEEE J. **45**(1), 235–248 (2010)
2. ATT: ATT u-verse-digital TV, high speed internet and voice (2012), http://www.att.com/shop/u-verse.html

3. P. Crombez, Full reconfigurable analog baseband circuits for multimode radios, KU Leuven Thesis (2009)
4. S. McGlaun, Tablet shipments to outpace total PC shipments in Q4 2013 says IDC, https://www.slashgear.com/tablet-shipments-to-outpace-total-pc-shipments-in-q4-2013-says-idc-12297574/
5. F. Wanlass, Low stand-by power complementary field effect circuitry, US Patent 3356858 (1967)
6. M. Gordon, Cramming more components onto integrated circuits, Electronics Magazine (1965)
7. W. Zhao, Y. Cao, New generation of predictive technology model for sub-45 nm early design exploration. IEEE Trans. Electron Devices **53**(11), 2816–2823 (2006)
8. W. Sansen, Analog design essentials (Springer, 2008)
9. K. Cornelissens, Delta-Sigma A/D converter design in nanoscale CMOS, KU Leuven Thesis (2010)
10. ITRS, International technology roadmap for semiconductors (2012), http://www.itrs.net/Links/2012ITRS/Home2012.htm,2012
11. J.L. Hogan, Developments of the heterodyne receiver. Proc. Inst. Radio Eng. **3**(3), 249–259 (1915)
12. A.A. Abidi, Direct-conversion radio transceivers for digital communications. IEEE J. Solid-State Circ. **30**(12), 1399–1410 (1995)
13. J. Crols, M.S.J. Steyaert, single-chip 900 MHz CMOS receiver front-end with a high performance low-IF topology. IEEE J. Solid-State Circuits **30**(12), 1483–1492 (1995)
14. T. Chalvatzis, T.O. Dickson, S.P. Voinigescu, 2-GHz direct-sampling delta-sigma tunable receiver with 40-GHz sampling clock and on-chip PLL, in 2007 IEEE Symposium on VLSI Circuits (2007), pp. 54–55
15. B. Grayver, E. Daneshrad, VLSI implementation of a 100-μw multirate FSK receiver. IEEE J. Solid-State Circuits, **36**(11), 1821–1828 (2001)

Chapter 2
A/D Converters and Applications

2.1 Introduction

Considering that a signal can be continuous or discrete in time and amplitude, there are four kinds of signals. Among these, the analog signal (continuous in time and amplitude) and the digital signal (discrete in time and amplitude) are widely used in electronic circuits and systems. Because of the nature of the CMOS transistor, it is a good switch component and a perfect choice for digital circuits (e.g. processor, ASIC, etc.), which is suitable for complex computations. However, all signals in the real world (e.g. sound, temperature, pressure, radio, etc.) are analog. As a result, data converters, i.e. ADC and digital-to-analog converters (DAC), are indispensable blocks for almost all electronic systems, behaving as bridges between the digital processing cores and the analog world. Although in some recent communication systems a power amplifier (PA) array is adopted to functionally combine a traditional PA and a DAC [1], ADCs (or its simplest structure: comparators) cannot be replaced or eliminated from systems.

In this chapter, first different specifications will be introduced as used to evaluate an ADC. Also two FoMs are given to compare ADCs with different architectures and specifications. For different applications and specifications, different ADC architectures are needed to meet the requirements. In the next section, the basic operation principles and the speed-performance limitations of different popular ADC architectures are discussed. A short comparison between different ADCs is also presented. Finally, we will give a brief introduction about the application of ADCs in communication systems, focusing on how to derive the block-level ADC accuracy requirement from the system-level receiver performance.

X. Xing et al., *Design of Power-Efficient Highly Digital Analog-to-Digital Converters for Next-Generation Wireless Communication Systems*, Signals and Communication Technology, https://doi.org/10.1007/978-3-319-66565-8_2

2.2 ADC Specifications

As basic blocks in electronic systems, ADCs are characterized by four parameters: speed, accuracy, power consumption and power efficiency, three of which are independent.

2.2.1 ADC Speed

The ADC speed defines how fast an ADC can convert its analog input signal into the corresponding digital signal. We distinguish the following terms.

- Bandwidth (BW):
  For a lowpass (or baseband) signal, the signal BW is its highest frequency; the BW of a bandpass (or passband) signal is its highest frequency minus its lowest frequency. The BW of an ADC is the maximum BW of the analog input beyond which the frequency aliasing phenomenon happens.
- Effective resolution bandwidth (ERBW):
  Normally the ADC accuracy is the highest for a low input signal frequency, and it drops gradually with increasing input frequency. The ADC ERBW is the frequency range where the ADC conversion accuracy is 3dB below its low-frequency value.
- Nyquist frequency (F_N):
  The ADC F_N is the minimum sampling rate required to prevent frequency aliasing. It is twice the ADC BW.
- Sampling frequency (F_S):
  F_S is the frequency at which the ADC converts the analog signal into a digital value. It is not directly related to the ADC bandwidth specifications. According to the Nyquist theorem, the F_S should be no smaller than the F_N. In Nyquist ADCs, the F_S is equal to the F_N; while in oversampling ADCs, the F_S is much larger than the F_N. In that case the oversampling ratio (OSR) is defined as the ratio of F_S and F_N:

$$OSR = \frac{F_S}{F_N} = \frac{F_S}{2 \cdot BW} \tag{2.1}$$

2.2.2 ADC Accuracy

The ADC accuracy specifications measure how large the difference is between the ADC analog input and the digital output. They are classified into static and dynamic parameters, depending on the ADC analog input signal frequency. The ADC static accuracy specifications are obtained in simulations or measurements for a direct-current (DC) signal. They are always expressed relative to the least significant bit (LSB).

- Gain error:
  The gain error of an ADC describes how accurate the slope of a real transfer function matches the slope of the ideal ADC transfer function.
- Offset:
  Because of ADC nonidealities, when the input is zero, the ADC output is not always zero. If a non-zero input voltage is needed to make the ADC output zero, this non-zero input is called the ADC offset.
- Differential nonlinearity error (DNL):
  Ideally, the difference between the analog voltages of two adjacent digital steps should be 1 LSB, which is not the case in real circuits. The ADC DNL is the largest difference between two adjacent analog voltage steps minus 1 LSB step.
- Integral nonlinearity error (INL):
  The ADC INL is the maximum analog deviation of the actual transfer function from the ideal ADC transfer function (after correcting for the offset and gain errors).

When the ADC analog input is an alternating-current (AC) signal, its accuracy is characterized by the dynamic specifications. All these specifications are calculated in the frequency domain and are expressed in dB (except for ENOB).

- Signal-to-noise ratio (SNR):
  The ADC SNR is the ratio between the signal power to the noise power integrated overall the whole signal bandwidth.
- Signal-to-noise-distortion ratio (SNDR):
  SNDR is the ratio between the signal power to the power of the integrated noise and the harmonic distortion components within the signal bandwidth.
- Effective number of bits (ENOB)
  A real ADC can be replaced by an ideal ADC with only quantization noise, whose SNR is the same as the SNDR of the real ADC. The number of bits of the ideal ADC is the ENOB of the evaluated ADC, it is a function of its SNDR:

$$ENOB = \frac{SNDR - 1.76}{6.02} \tag{2.2}$$

- Total harmonic distortion (THD):
  The THD is the ratio of the ADC signal power to the root-mean-square (RMS) power sum of all harmonic distortion components within the signal bandwidth.
- Dynamic range (DR):
  The ADC DR is defined by the ratio between the ADC's largest input signal power and the smallest input signal power. The largest input signal is the input when its SNDR drops by 3dB from the peak SNDR, and the minimum input signal is the same as the ADC noise floor.
- Spurious-free dynamic range (SFDR):
  The SFDR is the ratio between the fundamental signal power to the next largest spurious component within the signal bandwidth, excluding the DC offset.

- 2nd/3rd-order intermodulation distortion (IM2/IM3):
  In a two-tone test, if the test signal frequencies are f and $f + \Delta f$, then the 2nd-order intermodulation distortions are located at the frequencies Δf and $2f + \Delta f$, while the frequencies of the 3rd-order intermodulation distortions are $f - \Delta f, f + 2\Delta f$, $3f + \Delta f$ and $3f + 2\Delta f$. The ratio between the signal power (single-tone) to the 2nd- and 3rd-order intermodulation distortion powers are defined as IM2 and IM3 respectively.

Although several characteristics are described above to evaluate the ADC accuracy, in practical applications different ADC characteristics make sense and only some ADC accuracy specifications are important. Generally speaking, DR is important in sensor interfaces and instrumentation applications; THD is critical for audio systems; DNL is essential for video processing; and SFDR is the key accuracy specification for multi-channel radio communications.

2.2.3 *ADC FoM*

Different ADCs have different architectures and a different range of specifications. To make a comprehensive and fair comparison between these ADCs, some parameters combine the ADC speed, its accuracy and power consumption into a so-called figure-of-merit (FoM). In ADC design and applications, two FoMs [2, 3] are widely used:

$$FoM_1 = \frac{P}{2 \cdot BW \cdot 2^{ENOB}} = \frac{P}{2 \cdot BW \cdot 2^{\frac{SNDR-1.76}{6.02}}} \tag{2.3}$$

$$FoM_2 = SNDR + 10 \cdot log_{10}\left(\frac{BW}{P}\right) \tag{2.4}$$

where the P is the ADC power consumption. These two FoMs are expressed in pJ/Conversion (or pJ/Step) and dB respectively.

2.3 ADC Architectures

Generally speaking, a full A/D conversion can be splitted into four steps: anti-aliasing filtering, sampling and holding, quantization and encoding. In Fig. 2.1, a generalized ADC architecture with the basic blocks is shown. In this figure the signal waveforms and their corresponding spectrum plots for both Nyquist sampling and oversampling are illustrated as well.

- Anti-aliasing filter (AAF):
  According to the sampling theory, the sampled signal repeats its analog signal spectrum every multiple of the sampling frequency. The function of the AAF is to

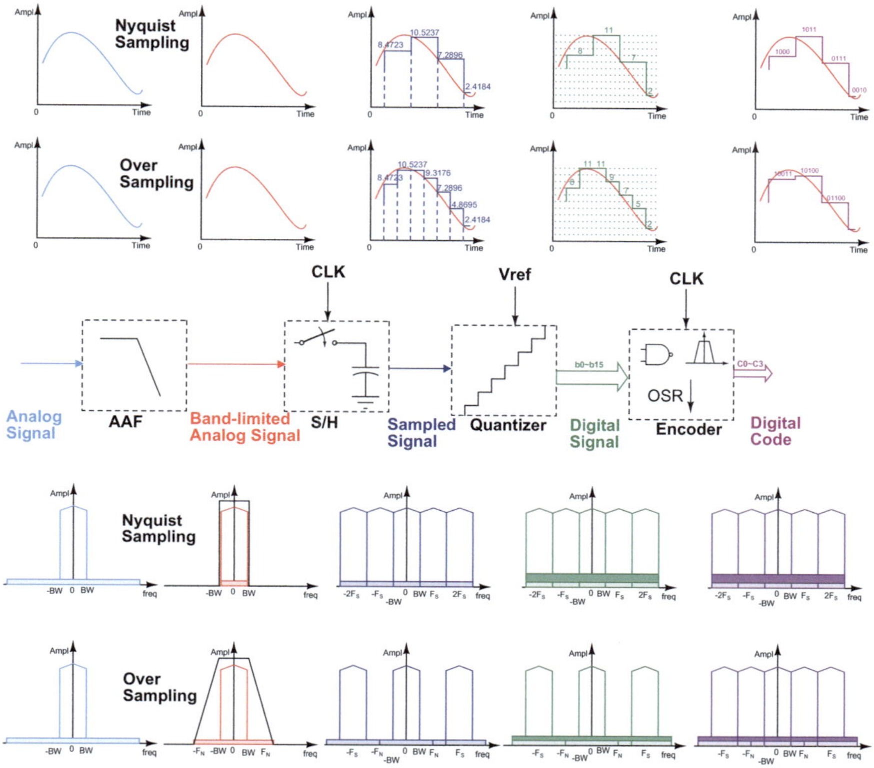

Fig. 2.1 Generalized architecture, operation and waveforms of an ADC, both with Nyquist sampling and oversampling

suppress the out-of-band noise and interference to a low enough level to prevent the desired signal from being polluted after being sampled. The different transfer functions of the AAF required in Nyquist sampling and oversampling are shown in Fig. 2.1.

- Sample-and-hold (S/H)
  The S/H is used to sample the analog input voltage and keep it constant during one conversion period, discretizing the signal in time. If the Nyquist theorem is met, the ideal sampling is a reversible process and the original analog signal can be extracted from the sampled signal by filtering. In a flash ADC architecture, the conversion is so fast that no S/H is typically needed. In ADC architectures with SC circuits (e.g. pipelined and discrete-time DSM ADCs), the S/H is combined with other circuits and no separate S/H is needed.
- Quantizer
  The sampled signal is now discrete in time, but still continuous in amplitude. The quantizer uses a limited set of preset values to approximate the infinite-accuracy input signal in amplitude, discretizing the sampled signal in amplitude. Due to the

quantization, a quantization error is added to the signal, which is not reversible. The quantizer block and its complexity vary largely among different ADC architectures, as shown in the following paragraphs. In the time domain, the quantization error is determined by the input voltage and the quantization levels (and therefore the number of bits), as shown in Fig. 2.2. If the quantizer input changes randomly from sample to sample, and it is not overloaded, the quantization error can be considered distributes uniformly over the interval $[-\frac{\Delta}{2},\frac{\Delta}{2}]$ and the total quantization error power is:

$$E_q^2 = \int_{-\infty}^{+\infty} e^2 pdf_e de = \frac{1}{\Delta}\int_{-\frac{\Delta}{2}}^{+\frac{\Delta}{2}} e^2 de = \frac{\Delta^2}{12} \quad (2.5)$$

In the frequency domain, this error distributes evenly over the range $[-\frac{F_S}{2},\frac{F_S}{2}]$ and the power spectral density is:

$$S_q^2(f) = E_q^2 \cdot \frac{1}{F_S} = \frac{\Delta^2}{12 \cdot F_S} \quad (2.6)$$

Without overloading, the quantizer maximum output signal power is:

$$P = \frac{1}{2}V_o^2 = \frac{1}{2}\left(\frac{(2^B-1)\Delta}{2}\right)^2 \approx \frac{2^{2B}\Delta^2}{8} \quad (2.7)$$

where B is the number of quantization bits, so the peak SNR is:

$$SNR_p = \frac{P}{E_q^2} = \frac{\frac{2^{2B}\Delta^2}{8}}{\frac{\Delta^2}{12}} = \frac{3}{2}\cdot 2^{2B} = 1.76 + 6.02B[dB] \quad (2.8)$$

By applying a larger F_S, the power spectral density of the quantization error is lower, and the quantization error integrated over the range $[-BW, BW]$ is reduced by the OSR, as presented in Fig. 2.1:

$$E_{q,osr}^2 = S_q^2(f)\cdot 2BW = \frac{\Delta^2 \cdot 2BW}{12 \cdot F_S} = \frac{\Delta^2}{12OSR} \quad (2.9)$$

So the peak SNR is improved by $OSR\times$:

$$SNR_{p,osr} = \frac{P}{E_{q,osr}^2} = \frac{3}{2}\cdot 2^{2B}OSR = 1.76 + 6.02B + 10log(OSR)[dB] \quad (2.10)$$

- Encoder
  The output of the quantizer is always thermometer code, which is not efficient for transmission. The encoder block transforms the quantizer output into the desired encoding. For oversampling ADCs, decimation is also needed, filtering out the noise beyond the signal bandwidth and reducing the data rate to F_N.

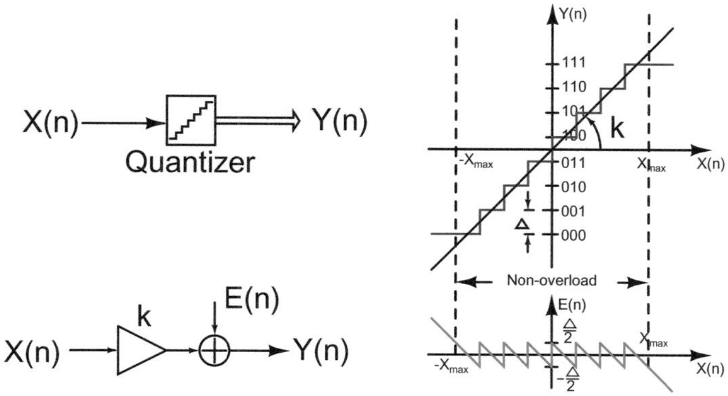

Fig. 2.2 The linear model and input-output transfer function of the quantizer

In this section, several popular ADC architectures will be introduced and discussed briefly, including Nyquist ADCs (flash, pipelined, two-step, SAR ADCs) and oversampling ADCs (DSM ADC). It should be noticed that Nyquist ADCs are also clocked in an oversampling way in most applications. There are two benefits from this arrangement. Firstly, the design of the anti-aliasing filter (AAF) preceding the ADC is relaxed. Furthermore, the ADC noise performance is improved by 3dB for every doubling of the clock frequency.

2.3.1 *Flash ADC*

The architecture of a N-bit flash ADC is illustrated in Fig. 2.3. It is composed of a resistor ladder, a pre-amplifier array, a comparator array and the digital processing part. The resistor ladder is used to generate a string of reference voltages with identical step size to bias the negative inputs of the pre-amplifier. The pre-amplifiers amplify the difference between the analog input voltage and the reference voltages, before driving the subsequent comparators. In some high-speed flash ADCs, more than one stage of pre-amplifier is used to speed up the conversion. The comparators make a logic decisions (0 and 1) based on the pre-amplifier outputs. In the digital part, NAND operations are introduced to remove code bubbles and a ROM is used for any thermometer-to-binary encoding.

The flash ADC sampling rate is limited by the speed of the pre-amplifier and the settling time of the comparators. It is the ADC topology with the highest signal bandwidth. A flash ADC with up to 40 GS/s sampling frequency in 0.12 μm SiGe technology is shown in [4]; the SNDR is 18.6 dB and the power is about 3.8 W. For the accuracy, flash ADCs highly depend on the transistor matching, which is determined by technology parameters and the transistor area. As a result, its chip area, the power consumption and the input capacitance are exponentially proportional to the number

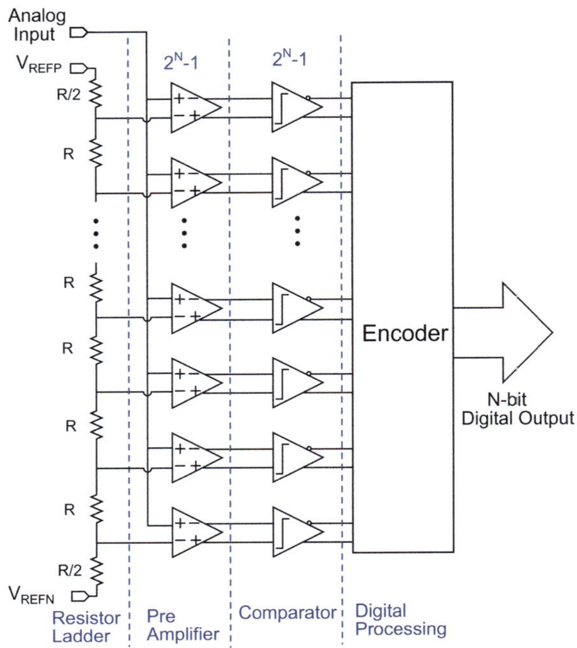

Fig. 2.3 N-bit flash ADC architecture

of bits. If the resolution is increased by 1-bit, the number of pre-amplifiers and comparators will double, and the area of each analog input transistor also needs to be doubled to improve the matching. So the chip area and the power consumption will be ×4. Moreover, the large input capacitance together with the output impedance of the preceding stage would reduce the maximum flash ADC bandwidth. As a result, flash ADC resolution is usually below 6 bit. So far, a 550 MHz-BW flash ADC with up to 35.8 dB SNDR has been reported [5], which dissipates about 300 mW power. In flash ADCs, due to the high throughput, a large portion of the power consumption comes from the digital part.

To solve the issues mentioned above, some techniques have been proposed and several new architectures have been invented. Analog interpolation is applied to reduce the number of pre-amplifiers, and folding is used to reduce the number of latch comparators. These two techniques can always be applied at the same time to increase the ADC resolution and/or to save flash ADC power consumption. For example, the SNR of a bipolar folding ADC reaches 75 dB for a 50 MHz signal BW with 1.25 W power consumption [6]. Limited by the folding circuit, the maximum sampling frequency of folding ADCs so far is about 2.2 GS/s, with 37.4 dB SNDR [7]. The FoM of this folding ADC is only 205.7fJ/Step, much better than typical value of flash ADCs (around 1 pJ/Step).

Averaging is another important innovation for flash ADCs [8], which is used to relax the transistor matching requirement, to reduce the chip area and to increase the ADC bandwidth. The averaging technique is based on the fact that deterministic signals are added in a linear way, while random signals (including mismatch and noise) are summed in a root-mean-square (RMS) manner.

$$SNR_{average} = 20log\frac{S_{average}}{E_{average}} = 20log\frac{\frac{S_1+S_2+\cdots+S_N}{N}}{\sqrt{\frac{E_1{}^2+E_2{}^2+\cdots+E_N{}^2}{N^2}}} \tag{2.11}$$

$$= 20log\frac{S_i}{\frac{E_i}{\sqrt{N}}} = 20log\frac{\sqrt{N}S_i}{E_i} \tag{2.12}$$

where $S_{average}$ and S_i are signal with and without averaging; $E_{average}$ and E_i are error after and before averaging. Here N comparator outputs are averaged, and the $SNDR$ is improved by $\sqrt{N}$.

2.3.2 *Two-Step ADC*

As mentioned in subsection 2.3.1, the chip area and the power consumption of a flash ADC will increase ×4 if its resolution is increased by 1 bit. One solution to this fast-increasing cost is to execute the conversion in two steps instead of single step, resulting into the two-step ADC architecture, as presented in Fig. 2.4. Due to the two-step conversion, its latency is larger than that of a flash ADC, so an S/H is needed to keep the analog input voltage constant during one conversion period. After being sampled and held, it works as follows. The input voltage is first converted by a M-bit coarse ADC; a M-bit DAC converts the digital bits back to the analog domain. Then the residual error is generated by subtracting the DAC output from the whole ADC input; the residual is amplified by 2^M according to the resolution of the coarse ADC and processed further by the N-bit fine ADC. The digital outputs of the coarse and fine ADCs are the MSB and LSB sets of the whole (M + N)-bit ADC digital output respectively. By doing the splitting, the hardware cost of a two-step ADC (including the extra hardware) is much smaller than that of a flash ADC with the same resolution:

$$2^M + 2^N << 2^{M+N} \tag{2.13}$$

The speed of a two-step ADC is determined by the delay of the critical path, which includes the S/H, the coarse ADC, the DAC, the subtractor, the amplifier and the fine ADC. Until now, a 1 GS/s two-step ADC with 33.8 dB SNDR has been presented, with about 55 mW power consumption [9]. Because of the preceding amplification, the fine-ADC accuracy requirement is relaxed. However, the accuracy requirements of the S/H, the coarse ADC, the DAC, the subtractor and the amplifier are the same as the whole ADC specification, which limits the two-step ADC resolution to about 10 bit. Also the matching between the ranges of the coarse and fine ADCs is critical to the overall ADC performance. One solution is assigning an "over-range" to the

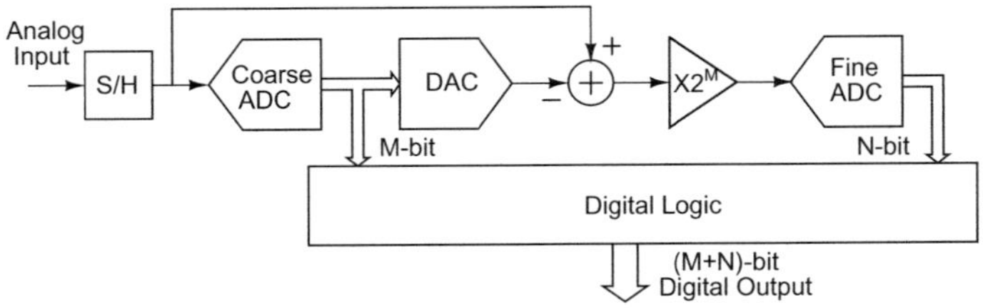

Fig. 2.4 Block diagram of a (M + N)-bit two-step ADC

fine ADC [10]. By addressing these bottlenecks, a 58.7 dB SNDR two-step ADC with 20 MS/s and 55 mW has been reported in [11].

2.3.3 Pipelined ADC

In a two-step ADC, if an additional S/H is inserted between the amplifier and the fine ADC and the number of stages is extended, then the pipelined ADC structure is generated, as shown in Fig. 2.5 (analog input buffer is not included). The analog part of a pipelined ADC is N cascaded conversion stages; all stages except the last stage (usually a flash ADC) are multiplying DACs (MDACs), as illustrated in the dashed box. Each MDAC first samples the input voltage, quantizes the sampled voltage, and then extracts the quantization error by using a DAC and a subtractor. Finally the residual quantization error is amplified and forwarded to the succeeding MDAC stage. All operations of the MDAC stage can be merged and executed by some SC circuit with one sampling phase and one amplifying phase. Because of the cascade structure, the accuracy requirement of each stage can be scaled, and full accuracy is only required in the first stage. Although the pipelined ADC can process a new sample in each clock cycle, the latency of a complete digital output is N clock periods, which limits the use of pipelined ADCs in control loops.

The simplest pipelined ADC architecture converts a single bit per stage. However, the gain error and offset of the MDAC stage influence the ADC performance directly. One solution to this problem is to use a 1.5-bit structure to compensate for the range mismatch [12]. Generally speaking, the bandwidth and performance of a pipelined ADC are mainly determined by the input buffer and the front-end MDAC OTAs, and these building blocks consume most of the power of the whole ADC. For a single-channel pipelined ADC, the maximum sampling frequency can reach 1 GS/s, with 69dB SNDR and 1.2 W power consumption [13]; with time interleaving, up to 20 GS/s has been achieved [14], with a SNDR of 29.5 dB and a power of 10 W. To reach high resolutions (>12bit), digital calibration is always needed. A 130 mW pipelined ADC with 84.9 dB SNDR for 2.5 MHz bandwidth has been presented [15].

To ease the ADC design and to optimize the power consumption, simple amplifiers or even comparators are used in the MDAC design [16], and any resulting error can

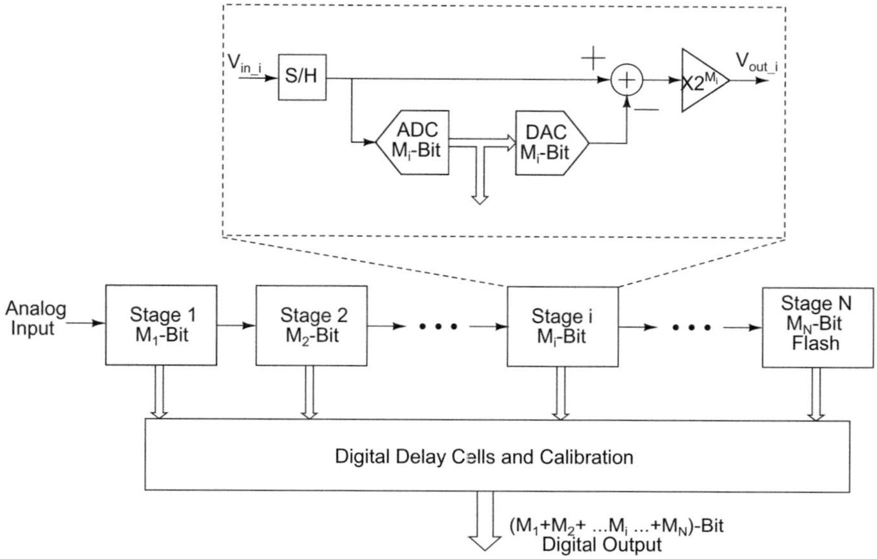

Fig. 2.5 Simplified architecture of pipelined ADC

be solved by digital calibration. Since there is no OTA used during the sampling phase, it can be shared between the sampling phase and the amplifying phase (of another MDAC stage), reducing the ADC power dissipation and chip area [17].

2.3.4 *SAR ADC*

The architecture of an N-bit SAR ADC is presented in Fig. 2.6. It consists of a S/H, a comparator, an N-bit DAC, a register and some control logic. The internal clock frequency (for the comparator, the DAC and the register) is N times the ADC frequency. At the beginning of each ADC clock period, the register bits are reset to zero, and the analog input voltage is sampled and held constant during the whole period. In the first internal clock cycle, the MSB of the register bits is preset to 1 and the DAC output is updated. Then the sampled input voltage is compared to the DAC output and the register MSB is kept to 1 or set back to 0, depending on the comparison result. In the next internal cycle, the MSB-1 bit is preset to 1 and the updated DAC output is again compared to the sampled input voltage. The comparison result decides the MSB-1 bit value. After N "preset-update-compare-determine" cycles in N internal clock periods, the LSB is determined and the complete N-bit conversion is done.

Because only one comparator and one DAC are analog in a SAR ADC, it can reach a very high power efficiency: a design with a FoM as low as 0.85fJ/Step have been reported [18]. In SAR ADCs, the performance directly depends on the linearity of the DAC, whose area always dominates the chip area. The nonidealities of the com-

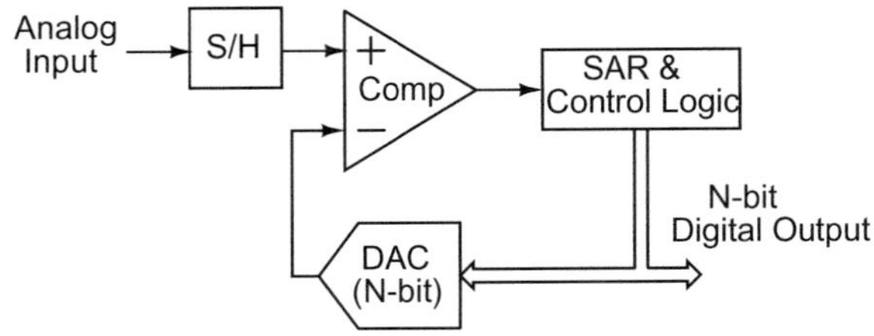

Fig. 2.6 Structure of SAR ADC

parator, e.g. the offset voltage and the kick-back noise are also critical to the overall performance. So far, an 12-bit ENOB SAR with 35 MS/s and 54.5 mW power dissipation has benn published [19]. The SAR ADC's maximum frequency is determined by its resolution and the total delay of the comparator and the DAC. With a single channel, SAR ADCs can reach 1.2 GS/s maximum sampling frequency with about 39.3 dB SNDR and 3.1 mW power [20]. With 80 channels time interleaving, the clock frequency impressively reaches 56 GS/s, with 5.7bit ENOB and 2 W power in 65 nm CMOS [21]. However, because of the additional power from the clock generation circuit and the multiplexing, the ADC FoM is 0.69pJ/Step, much larger compared to single-channel SAR ADCs. By using time interleaving and oversampling, the resolution of a SAR ADC can reach 14 bit [22].

2.3.5 Delta-Sigma ADC

Until now all ADC architectures introduced above in this section are Nyquist ADCs. In theory their sampling rate can be as low as the Nyquist frequency. Contrary to Nyquist ADCs, the sampling frequency of a Delta-Sigma (DSM) ADC is much higher than its Nyquist frequency. When doubling the sampling rate of a Nyquist ADC, the SNR is improved by 3 dB, so the accuracy-speed exchange efficiency is still too low. One solution to this problem is the noise shaping technique, which is always done by negative feedback. The combination of oversampling and noise shaping results in the DSM ADC, which is shown in Fig. 2.7. The DSM core is a negative feedback loop with both analog and digital signals. The loop filter compares the analog input and the feedback DAC's output and amplifies/integrates the difference. The filter output goes into the quantizer and the digital output stream is obtained. Because of the global negative feedback and the high gain of the loop filter within the signal bandwidth, the M-bit digital stream tracks the analog input on average. The quantization noise is not only oversampled, but also partly pushed from the band of interest to the band beyond. In the digital decimation circuit, the out-of-band quantization noise of the digital stream is suppressed, and the data rate is then reduced to the Nyquist rate by a sinc filter.

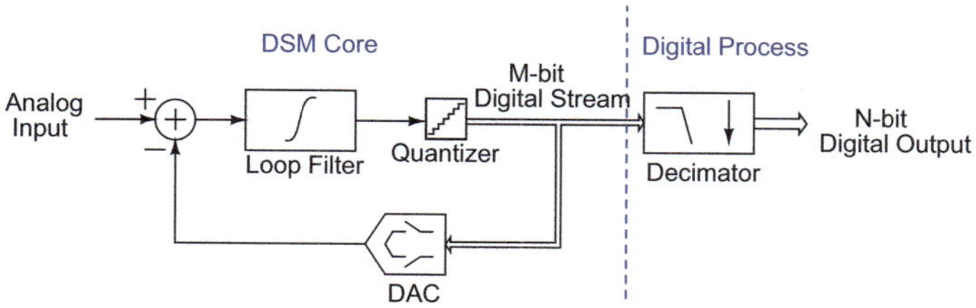

Fig. 2.7 Block diagram of DSM ADC

In a DSM ADC, the performance is determined by three parameters: the number of quantization bits B, the oversampling ratio OSR and the noise shaping order N [3]:

$$SNR = 6.02B + 1.76 + 10(2N + 1)log(OSR) - 10log\left(\frac{\pi^{2N}}{2N + 1}\right) \tag{2.14}$$

Generally, high-accuracy DSM ADCs are implemented by discrete-time (DT) DSMs, which are widely used in high-quality audio systems. The performance of DT DSMs is limited by the settling accuracy of the integrators in the loop filter. So far, a 116 dB SNDR 1 kHz-BW ADC has been presented in [23], with 12.7 mW power. The highest BW of a DT DSM ADC is 20 MHz, with 64dB SNDR and 27.9 mW power consumption [24]. With relaxed settling requirements on the loop filter, continuous-time (CT) DSM ADCs have extended the bandwidth to tens of MHz (with lower accuracy) and found application in communications. In such ADCs, the bandwidth is limited by the speed of the loop filter and the multi-bit quantizer. For example, a 465 MHz signal bandwidth CT DSM ADC has been reported, with 64.7 dB SNDR and 930 mW power consumption [25]. With 24 kHz and 280 μW power, a CT DSM ADC has reached a highest SNDR of 98.5 dB [26].

Compared to Nyquist ADCs, one advantage of DSM ADCs is the relaxed (e.g. simple RC filter) or even eliminated anti-aliasing filter, especially for CT DSM ADCs. DSM ADCs need one extra digital decimation filter, however, instead of an analog filter; it is robust and easy to design, furthermore its area and power dissipation can be reduced with CMOS technology scaling. In DSM ADCs, different choices of each block will lead to different structures and they all show benefits and limitations. Since the digital part can be synthesized by a digital design flow using standard libraries from the foundry, it is not covered in this work. The book is about the design consideration of the DSM core (closed-loop ones and open-loop ones), which will be presented in the following chapters.

2.3.6 ADC Architecture Summary and Comparison

As a summary of this subsection, the performance limitations, the time needed for one complete conversion and the hardware cost of the five discussed ADC topologies are listed in Table 2.1. Here N stands for the ADC resolution and T for the conversion period of a basic comparator. In Fig. 2.8, the achievable ENOB and bandwidth of each ADC in today's CMOS technology are shown. It is observed that for different ADC architectures and for each architecture itself, there are tradeoffs between bandwidth and accuracy, and the choice among these ADC architectures highly depends on the application and its requirement. With CMOS technology scaling and circuit design innovations, both the bandwidth and the ENOB range of each ADC type will become larger and there will be more overlap between different ADC topologies. In that case, the power efficiency, chip area, design complexity and robustness will be the selection criteria.

To give a general impression about power efficiencies of different ADC architectures, a ADC FoM plot based on ISSCC publications [27] is presented in Fig. 2.9. Here several phenomena and trends can be observed.

- The FoMs of SAR, pipelined and DSM (both DT and CT) ADCs improve by around 1.5× every year, but the FoM improvements of flash and two-step ADCs are much slower.
- Besides flash ADC, all other ADC architectures can reach a FoM smaller than 100fJ/Step, which is a consequence of technology advancements and circuit design innovations. Only pipelined and SAR ADCs can achieve a FoM better than 10fJ/Step, and these two topologies and their time-interleaving counterparts are popular recently. Among all topologies, SAR ADC has the lowest FoM, and it is the only choice for a FoM lower than 1fJ/Step.

Table 2.1 Performance summary of different ADC architectures

ADC architecture	Accuracy limitation	Clock cycles	Hardware cost
Flash	Transistor intrinsic matching, kick-back noise	T	$\propto 2^N$
Two-step	coarse and fine stages range matching	$2T$	$\propto 2^{\frac{N+2}{2}}$
Pipelined (M stages)	MDAC gain and offset	$M \cdot T$	$\propto M \cdot 2^{\frac{N}{M}}$
SAR	Feedback DAC matching, comparator offset and kick-back noise	$N \cdot T$	$\propto N$
DSM (2nd-order, 1-bit)	Integrator settling	$2^{0.4N+1} \cdot T$	–

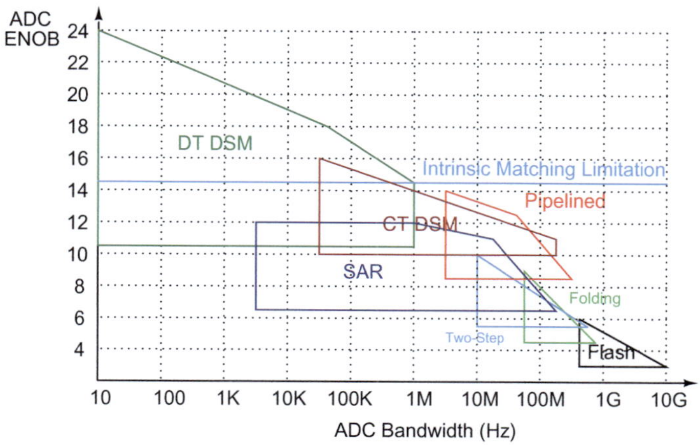

Fig. 2.8 Accuracy-bandwidth tradeoff of ADC architectures

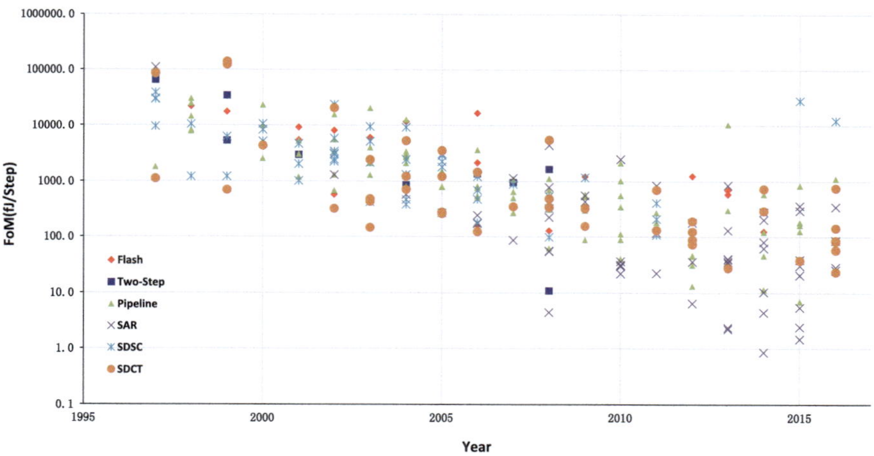

Fig. 2.9 FoMs of different ADC architectures in ISSCC publications [27]

- Both pipelined and CT DSM ADCs are qualified for communication applications with less than 100 MHz bandwidth, and they have almost the same FoM. However, considering the large power consumed by the AAF preceding a pipelined ADC, CT DSM ADCs will replace more and more pipelined ADCs in communication terminal applications in the future.

2.4 Application of ADC in Communications

Every complete receiver can be divided into three parts: the analog front end (AFE), the ADC and the DSP, as shown in Fig. 2.10. All circuits preceding the ADC are grouped into the AFE part, consisting out of the RF circuits, IF circuits, baseband (BB) circuits and mixers. Different types of AFE have been presented and discussed in Chap. 1, and a typical receiver AFE is shown in 2.11. The function of the AFE is to perform frequency downconversion, and the necessary filtering and amplification. It precondition the antenna output into a stable signal which can be digitized by the ADC. The DSP is used to do demodulation to recover the information data. The ADC in between converts the signal from the analog domain to the digital domain and behaves as the unidirectional signal bridge between the AFE and the DSP. Today it is the performance bottleneck in more and more receiver designs [28].

In receiver system-level design, the receiver topology determines the optimized ADC architecture. Two lowpass ADCs with quadrature clocks are suitable in a zero-IF receiver. On the other hand, in a low-IF receiver, a quadrature ADC is preferred to avoid the DC offset problem. In heterodyne receivers, a lowpass ADC is enough for a BB conversion while a bandpass ADC is optimum in power for a IF/RF conversion. For the ADC bandwidth requirement, it is mainly determined by the communication channel bandwidth, the number of channels converted by the ADC and the amount of oversampling. The accuracy demanded from an ADC is however based on more factors and the derivation is much more complex, as will be covered in this section.

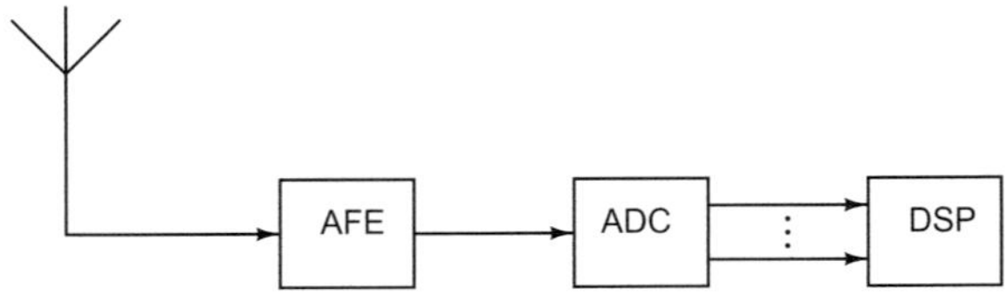

Fig. 2.10 A receiver is partitioned into three basic parts

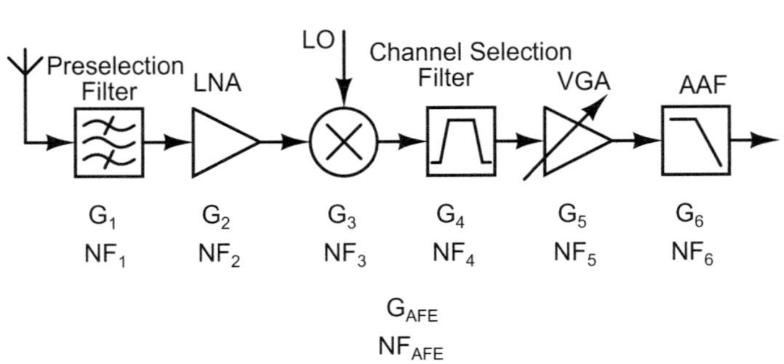

Fig. 2.11 Receiver AFE is a cascaded system

First, we treat the AFE as a whole entity and calculate its gain (G_{AFE}) and noise figure (NF_{AFE}), using the classic Friis formulas for cascaded stages:

$$G_{AFE}(dB) = G_1 + G_2 + \cdots + G_5 + G_6 \tag{2.15}$$

$$F_i = 10^{\frac{NF_i}{10}} \tag{2.16}$$

$$F_{AFE} = F_1 + \frac{F_2 - 1}{G_1} + \frac{F_3 - 1}{G_1 G_2} + \cdots + \frac{F_6 - 1}{G_1 G_2 G_3 G_4 G_5} \tag{2.17}$$

$$NF_{AFE} = 10log(F_{AFE}) \tag{2.18}$$

where G_i, F_i and NF_i are the gain, noise factor and noise figure of the ith stages respectively.

In general, the ADC accuracy must be high enough so that it can handle both the smallest and the largest AFE outputs in the receiver. Also the A/D conversion may not cause too much accuracy degradation in the signal chain. The adopted analysis aims to find the highest power level (P_{max}) and the lowest power level (P_{min}) at the ADC input, and the required ADC accuracy is the difference between them:

$$SNDR(dB) = P_{max} - P_{min} \tag{2.19}$$

The highest power level (P_{max}) is always found when the receiver receives the highest signal power as determined by several factors listed below [29].

- The received maximum signal power (R_{max})
  In communication standards, to avoid too much inteference to other communication systems, the power received by the antenna should not exceed the defined maximum power (R_{max}).
- The AFE minimum gain (G_{AFE_min})
  To accommodate to the high input power from the free space, the AFE gain is set to its lowest value G_{AFE_min} to avoid saturating the ADC.
- The crest factor (CF)
  In ADC evaluation, the SNDR is obtained with a peak signal amplitude, while in many communication applications the RMS value of the carrier waveform does not reach this peak value. This carrier peak-to-rms ratio is also called the crest factor (CF). The CF value depends on the modulation. For example, the CF of 64 QAM is 3.7 dB [28].
- DC offset ($OFST$)
  In some receiver architectures, because of VCO leakage and mixer self-mixing, there is a DC offset ($OFST$) with the AFE output signal, which is larger than the desired channel amplitude. To make sure that the receiver works properly, the DC offset should not saturate the ADC.

- Blocker (*BLKR*)
  After filtering and suppression, there may still be some residual blocker signal (*BLKR*) left, which is nearby the desired channel in spectrum. This blocker should also not affect the A/D conversion of the signal channel.
- Fading Margin (*FM*)
  Because of the complex propagation environment, some margin is needed for the channel signal to allow some fading, which is called the fading margin (*FM*). A typical FM for a GSM system is 3 dB.

If we take all these factors into consideration, then the ADC's highest power level (P_{max}) can be expressed as:

$$P_{max}(dB) = R_{max} + G_{AFE\_min} + CF + max\{OFST, BLKR\} + FM \tag{2.20}$$

The ADC lowest power level is decided by the ADC's input-referred noise, which can be calculated based on the following terms.

- The receiver sensitivity (R_{ss})
  In the AFE input, the received channel noise N_{in} is proportional to the channel bandwidth BW_{RF}:
  $$N_{in} = kTBW_{RF} \tag{2.21}$$
  where k is the Boltzman constant and T is the absolute temperature. There is a minimum receiver input signal power needed to provide an adequate accuracy for the receiver to do the demodulation correctly, which is called the receiver sensitivity (R_{ss}). Like the received maximum power R_{max}, the required receiver sensitivity R_{ss} is also defined in the communication standard.
- The required SNR (SNR_{req})
  The difference between the receiver sensitivity and the total channel noise (sum of the received channel noise and the AFE thermal noise) is defined as the required SNR SNR_{req}. The SNR_{req} value is a function of the signal modulation and the required bit error rate (BER_{req}). With a larger SNR_{req}, the channel capacity will increase and the BER can be reduced. For example, in 16QAM, a BER_{req} of 1e − 9 results in a SNR_{req} of 35dB. With this relationship, the noise figure of the AFE is decided:
  $$NF_{AFE\_max} = R_{ss} - kTBW_{RF} - SNR_{req} \tag{2.22}$$
- The AFE maximum gain (G_{AFE_max})
  With the minimum input signal, the AFE gain is set to its maximum value (G_{AFE_max}) to avoid the signal from being polluted by ADC noise.

- The ADC noise margin (NM_{ADC})
  For the ADC, the noise of the analog input is the received channel noise amplified by the maximum AFE gain, plus the AFE noise figure:

$$\begin{aligned} N_{ch} &= kTB_{RF} + G_{AFE\_max} + NF_{AFE\_max} \\ &= R_{ss} + G_{AFE\_max} - SNR_{req} \end{aligned} \tag{2.23}$$

  The ratio between the ADC input noise (channel noise) and the ADC noise (quantization noise, thermal noise and clock jitter noise) is called the ADC noise margin, this margin must be large enough so that the ADC noise does not degrade the receiver noise figure too much (normally less than 1 dB). A typical value of the ADC noise margin in wideband communication systems is 17 dB.

$$N_{ADC} = N_{ch} - NM_{ADC} \tag{2.24}$$

$$\begin{aligned} \Delta NF &= NF_{AFE\_ADC\_max} - NF_{AFE\_max} \\ &= 10log[10^{\frac{N_{ch}}{10}} + 10^{\frac{N_{ADC}}{10}}] - N_{CH} - NF_{AFE\_max} \\ &= 10log[10^{\frac{kTBW_{RF}+G_{AFE\_max}+NF_{AFE\_max}}{10}} \\ &+ 10^{\frac{kTBW_{RF}+G_{AFE\_max}+NF_{AFE\_max}-NM_{ADC}}{10}}] \\ &- kTBW_{RF} - G_{AFE\_max} - NF_{AFE\_max} < 1dB \end{aligned} \tag{2.25}$$

Considering all the above mentioned factors, the ADC lowest power level is given by:

$$P_{min}(dB) = R_{ss} + G_{AFE\_max} - SNR_{req} - NM_{ADC} \tag{2.26}$$

With the ADC maximum and minimum input power levels, the ADC's required SNDR can be calculated with the following formula. The complete ADC SNDR deduction procedure is graphically illustrated in Fig. 2.12.

$$\begin{aligned} SNDR(dB) = R_{max} &+ G_{AFE\_min} + CF \\ &+ max\{OFST, BLKR\} + FM \\ &- R_{ss} - G_{AFE\_max} \\ &+ SNR_{req} + NM_{ADC} \end{aligned} \tag{2.27}$$

Here we can define the difference between the received maximum power R_{max} and the receiver sensitivity R_{ss} as the receivable signal range (RSR), and the difference between the AFE maximum and minimum gains as the VGA gain range (GR_{VGA}).

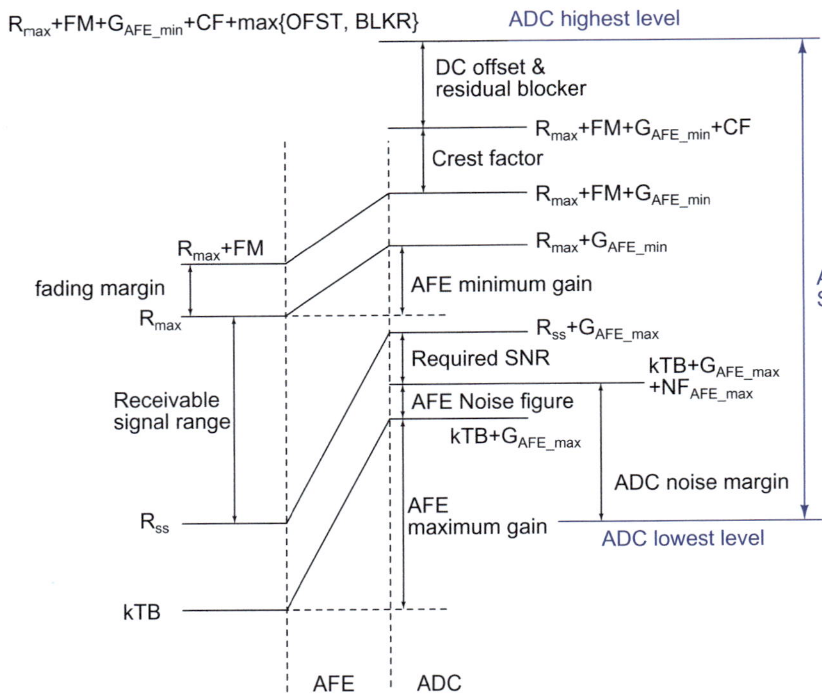

Fig. 2.12 Accuracy relationship between the ADC and the AFE in a telecom receiver

$$RSR = R_{max} - R_{ss} \tag{2.28}$$

$$GR_{VGA} = G_{AFE\_max} - G_{AFE\_min} \tag{2.29}$$

$$\begin{aligned} SNDR(dB) = & RSR - GR_{VGA} + CF \\ & + max\{OFST, BLKR\} + FM \\ & + SNR_{req} + NM_{ADC} \\ & - 10log10(BW_{ADC}/BW_{RF}) \end{aligned} \tag{2.30}$$

It can be seen that if the communication channel is oversampled by the ADC, the ADC SNDR can be relaxed. To summarize, all factors influencing the required ADC SNDR in a wireless receiver are as follows:

- the receivable signal range (RSR)
- the VGA gain range (GR_{VGA})
- the crest factor (CF)
- the DC offset ($OFST$)
- the blocker ($BLKR$)

- the fading Margin (FM)
- the required SNR (SNR_{req})
- the ADC noise margin (NM_{ADC})
- the oversampling ratio (OSR)

In certain receiver architectures, some factors mentioned above do not exist, which can be set to 0dB during the calculation. With formula 2.30, some tradeoffs between the AFE and ADC design are shown clearly. For example, with a larger VGA gain range and a better suppression of blockers in the AFE, a smaller ADC SNDR is needed and the ADC power consumption can be reduced. On the other hand, with a larger ADC SNDR, the VGA and filter can be relaxed or even eliminated.

2.5 Conclusions

In this chapter, the ADC and its application in communication system have been discussed briefly. First, several specifications have been presented to evaluate the ADC in speed and accuracy. Two FoMs have also been introduced to compare different ADCs. In practical systems, the relevant accuracy specifications of interest are different.

For different specifications and applications, different ADC architectures are needed, and have been described. The flash ADC is the topology achieving the highest sampling frequency and with the lowest resolution. The two-step ADC splits the A/D conversion into two steps and can achieve a higher resolution, but its speed is reduced compared to the flash ADC. The SAR ADC is based on a binary search algorithm, and is known as the topology with the best power efficiency. The pipelined ADC is the unique architecture which can achieve high sampling frequency and high resolution simultaneously, and it is the most frequently used ADC structure in communications. Normally either high-performance OTA or digital calibration are required in a pipelined ADC. The DSM ADC is based on oversampling and noise shaping techniques. Because of the intrinsic AAF, more and more applications shift from Nyquist ADCs to DSM ADCs, especially CT DSMs.

In a receiver, the design of the ADC and the AFE depends on each other. The ADC must be able to handle the AFE output signal without causing too much accuracy loss. The full procedure of how to determine the ADC accuracy requirement has been illustrated, and all effecting factors have been listed. With the tradeoffs expressed by the resulting equation, a system-level optimization between the AFE, the ADC and even the signal modulation can be done. The next chapter will discuss about the design of closed-loop DSM ADCs.

References

1. M. Park, M.H. Perrott, R.B. Staszewski, A time-domain resolution improvement of an RF-DAC. IEEE Trans. Circuits Syst. II Express Briefs **57**(7), 517–521 (2010). July
2. R.H. Walden, Analog-to-digital converter survey and analysis. IEEE J. Sel. Areas Commun. **17**(4), 539–550 (1999). Apr
3. R. Schreier, G.C. Temes, *Understanding Delta-Sigma Data Converters* (Wiley, New York, 2005)
4. W. Cheng, W. Ali, Moon-Jung Choi, K. Liu, T. Tat, D. Devendorf,L. Linder, R. Stevens, A 3b 40GS/s ADC-DAC in 0.12μm SiGe, in 2004 IEEE InternationalSolid-State Circuits Conference, Digest of Technical Papers (ISSCC) (2004), pp. 262–263
5. G. Geelen, 6b 1.1GSample/s CMOS A/D converter, in 2001 IEEE International Solid-State Circuits Conference, Digest of Technical Papers (ISSCC) (2001), pp. 128–129
6. C. Moreland, M. Elliott, F. Murden, J. Young, M. Hensley, R. Stop, A 14b 100MSample/s 3-stage A/D converter, in 2000 IEEE International Solid-State Circuits Conference, Digest of Technical Papers (ISSCC) (2000), pp. 34–35
7. M. Miyahara, I. Mano, M. Nakayama, K. Okada, A. Matsuzawa, 22.6 A 2.2GS, s 7b 27.4 mW time-based folding-flash ADC with resistively averaged voltage-to-time amplifiers, in IEEE International Solid-State Circuits Conference Digest of Technical Papers (ISSCC). San Francisco, CA (2014), pp. 388–389
8. K. Kattmann, J. Barrow, A technique for reducing differential non-linearity errors in flash A/D converters, in Solid-State Circuits Conference, Digest of Technical Papers (1991), pp. 170–171
9. P. M. Figueiredo, P. Cardoso, A. Lopes, C. Fachada, N. Hamanishi,K. Tanabe, J. Vital, A 90 nm CMOS 1.2 V 6b 1GS/s two-step subranging ADC, in Solid-State Circuits Conference, ISSCC 2006. Digest of Technical Papers. IEEE International (2006), pp. 2320–2329
10. S.H. Lewis, P.R. Gray, A pipelined 5-Msample/s 9-bit analog-to-digital converter. IEEE J. Solid-State Circ. **22**(6), 954–961 (1987). June
11. B. Brandt, J. Lutsky, A 75-mW 10-b 20-MSample/s CMOS subranging ADC with 59-dB SNDR, in 1999 IEEE International Solid-State Circuits Conference, Digest of Technical Papers (ISSCC) (1999), pp. 322-323
12. X. Wang, P.J. Hurst, S.H. Lewis, A 12-bit 20-MS/s pipelined ADC with nested digital background calibration, in Proceedings of IEEE Custom Integrated Circuits Conference (2003), pp. 409–412
13. M.A. Ahmed, Ali, H. Dinc, P. Bhoraskar, C. Dillon, S. Puckett, B. Gray, C. Speir, J. Lanford, J. Brunsilius, P.R. Derounian, B. Jeffries, U. Mehta, M. McShea, R. Stop, A 14 bit 1 GS/s RF sampling pipelined ADC with background calibration. IEEE J. Solid-State Circ. **49**(12), 2857–2867 (2014)
14. K. Poulton, R. Neff, B. Setterberg, B. Wuppermann, T. Kopley, R. Jewett, J. Pernillo, C. Tan, A. Montijo, A 20 GS/s 8 b ADC with a 1 MB memory in 0.18 μm CMOS, in 2003 IEEE International, Solid-State Circuits Conference, Digest of Technical Papers, ISSCC (2003), pp. 318–319
15. Sung-Ung Kwak, Bang-Sup Song, K. Bacrania, A 15b 5MSample/s low-spurious CMOS ADC, in 1997 IEEE International Solid-State Circuits Conference, Digest of Technical Papers. 43rd ISSCC (1997), pp. 146–147
16. L. Brooks, Hae-Seung Lee, A 12b, 50 MS/s. Fully differential zero-crossing based pipelined ADC. IEEE J. Solid-State Circ. **44**(12), 3329–3343 (2009)
17. N. Sasidhar, Youn-Jae Kook, S. Takeuchi, K. Hamashita, K. Takasuka, P.K. Hanumolu, Un-Ku Moon, A low power pipelined ADC using capacitor and opamp sharing technique with a scheme to cancel the effect of signal dependent kickback. IEEE J. Solid-State Circ. **44**(9) 2392–2401 (2009)
18. H.Y. Tai, Y.S. Hu, H.W. Chen, H.S. Chen, 11.2 A 0.85fJ, conversion-step 10b 200kS, s subranging SAR ADC in 40 nm CMOS, IEEE International Solid-State Circuits Conference Digest of Technical Papers (ISSCC). San Francisco, CA (2014), pp. 196–197

19. M. Krmer, E. Janssen, K. Doris, B. Murmann, 14b 35MS, S SAR ADC achieving 75dB SNDR and 99dB SFDR with loop-embedded input buffer in 40nm CMOS, in IEEE International Solid-State Circuits Conference – (ISSCC) Digest of Technical Papers. San Francisco, CA (2015), pp. 1–3
20. L. Kull1, T. Toifl, M. Schmatz, P. A. Francese, C. Menolfi, M. Braendli, M. Kossel, Thomas Morf, T. M. Andersen, Y. Leblebici, A 3.1 mW 8b 1.2 GS/s single-channel asynchronous SAR ADC with alternate comparators for enhanced speed in 32 nm digital SOI CMOS, 2013 IEEE International Solid-State Circuits Conference Digest of Technical Papers, San Francisco (2013)
21. I. Dedic, 56Gs/s ADC: Enabling 100GbE, 2010 Conference on Optical Fiber Communication (OFC/NFOEC), collocated National Fiber Optic Engineers Conference, San Diego, CA, (2010), pp. 1–3
22. M. Hesener, T. Eichler, A. Hanneberg, D. Herbison, F. Kuttner, and H. Wenske, A 14b 40MS/s redundant SAR ADC with 480MHz clock in 0.13μm CMOS, in IEEE International Solid-State Circuits Conference, ISSCC 2007, Digest of Technical Papers, pp. 248–249 (2007)
23. M. Steiner, N. Greer, A 22.3b 1 kHz 12.7 mW switched-capacitor $\Delta\Sigma$ modulator with stacked split-steering amplifiers, IEEE International Solid-State Circuits Conference (ISSCC). San Francisco, CA (2016), pp. 284–286
24. P. Malla, H. Lakdawala, K. Kornegay, and K. Soumyanath, A 28mW spectrum-sensing reconfigurable 20MHz 72dB-SNR 70dB-SNDR DT ADC for 802.11n/WiMAX receivers, in Solid-State Circuits Conference, 2008. ISSCC 2008. Digest of Technical Papers. IEEE International, pp. 496-631, Feb. 2008
25. Y. Dong, J. Zhao, W. Yang, T. Caldwell, H. Shibata, R. Schreier, Q. Meng, J. Silva, D. Paterson, J. Gealow, 15.5 A 930 mW 69 dB-DR 465 MHz-BW CT 1–2 MASH ADC in 28 nm CMOS, IEEE International Solid-State Circuits Conference (ISSCC). San Francisco, CA (2016), pp. 278–279
26. S. Billa, A. Sukumaran, S. Pavan, A 280 μW 24 kHz-BW 98.5 dB-SNDR chopped single-bit CT $\Delta\Sigma$M achieving $<$10 Hz 1/f noise corner without chopping artifacts, in Solid-State Circuits Conference, Digest of Technical Papers, ISSCC, pp. 276–277 (2016)
27. B. Murmann, ADC Performance Survey 1997–2016 (2017), http://www.stanford.edu/murmann/adcsurvey.html
28. J. Mikael Gustavsson, *Jacob Wikner, and Nianxiong Tan* (Springer, CMOS data converters for communications, 2000)
29. Q. Gu, RF system design of transceivers for wireless communications (Springer, 2005)

Chapter 3
Continuous-Time Delta-Sigma Modulators

3.1 Introduction

DSM ADCs trade off speed for accuracy by using oversampling and noise shaping techniques. Compared to discrete-time (DT) DSMs, usually continuous-time (CT) DSMs are applied for higher-bandwidth and lower-accuracy A/D conversions. Three system-level parameters determine the ideal DSM ADC performance: the number of quantization bits (B), the oversampling ratio (OSR) and the noise shaping order (N). For the quantizer, both single-bit and multi-bit have their advantages and disadvantages. Generally the noise shaping order should not be higher than 5th (lowpass case) in a single-loop structure for modulator stability. The multi-stage noise shaping (MASH) topology increases the shaping order and maintains the stability simultaneously. However, a noise leakage issue comes from the analog/digital mismatch. In single-loop DSMs, feedforward, feedback and hybrid structures have been developed to accommodate different design requirements. In circuit design, both the active-RC and the Gm-C integrators can be used for loop filter realization. Current, voltage and charge (switched-capacitor) type DACs provide different choices for the feedback DACs. DSMs with different combinations of these parameters have different characteristics and trade off in different aspects (noise, distortion, power, bandwidth, clock jitter, etc.). It is important to select the most suitable structure based on the exact specification requirements and condition limitations. For each block in a CT DSM, different nonidealities have different mechanisms and effects on the modulator performance, such as the amplifier's limited gain and GBW, DAC mismatch, clock jitter, quantizer delay and so on. The effect of the nonideality also depends on the modulator structure. To quickly predict the modulator performance and to determine the design requirements for each building block, all important nonidealities need to be modelled correctly in behavioral-level simulations in Matlab.

This chapter covers the design of CT DSM ADCs. Firstly, the DSM basic operation principle is presented. Then, comparisons between different DSM topologies are given to help choose the optimized structure. Finally, the mechanisms and modelling

X. Xing et al., *Design of Power-Efficient Highly Digital Analog-to-Digital Converters for Next-Generation Wireless Communication Systems*, Signals and Communication Technology, https://doi.org/10.1007/978-3-319-66565-8_3

of important building block nonidealities in CT DSM are presented; also potential solutions for these nonidealities are discussed. Based on these contents, a complete CT DSM design flow can be summarized.

3.2 DSM Basics: Oversampling and Noise-Shaping

Being different from all other Nyquist ADC architectures, a DSM ADC has a sampling frequency much higher than its Nyquist frequency, so it is also named as oversampling ADC. A whole DSM ADC structure is presented in Fig. 3.1; it is composed of an AAF filter, a DSM core and a decimation filter.

The AAF filter is applied for out-of-band noise and distortion suppressing to avoid aliasing phenomena during sampling. Since in DSM ADCs the AAF filter transition band is large, the filter design is relaxed. In CT DSM ADCs, with the intrinsic AAF of the CT loop filter, only a simple RC filter is required as explicit AAF, saving considerable power consumption. The DSM core is the interface circuit which converts the analog signal into digital bits at the sampling frequency and moves part of the in-band quantization noise beyond the signal bandwidth. In the DSM core output, the digital bit rate is much higher than the Nyquist frequency and the out-of-band noise is high. A digital decimation filter (always implemented as a sinc filter) is needed after the A/D conversion to filter out the out-of-band noise and reduce the data rate back to the Nyquist frequency.

As mentioned previously, doubling the ADC sampling frequency will improve the SNR by 3dB. To improve the speed-accuracy tradeoff, the noise-shaping technique is introduced and combined together with the oversampling technique, resulting into a DSM [1]. A generalized DSM is illustrated in the top of Fig. 3.2, which is composed of a loop filter, a quantizer and a feedback DAC. Here the loop filter is a linear block and the feedback DAC can be considered as a path with unit gain. To analyse the DSM in the frequency domain, the quantizer can also be modelled as a linear block as long as its input is random and it is not overloaded [2]. In this model, if we assume that the loop filter is a N-stage cascaded integrator and the gain of the quantizer k is unity, then the DSM output Y can be expressed as a function of the input X and the quantization noise E:

$$Y = \frac{H(s)k}{1 + H(s)k} X + \frac{1}{1 + H(s)k} E = STF(s) \cdot X + NTF(s) \cdot E \tag{3.1}$$

And the signal transfer function (STF) and the noise transfer function (NTF) are respectively:

$$STF(s) = \frac{f_s^N}{s^N + f_s^N} \approx 1; NTF(s) = \frac{s^N}{s^N + f_s^N} \approx \frac{s^N}{f_s^N} \tag{3.2}$$

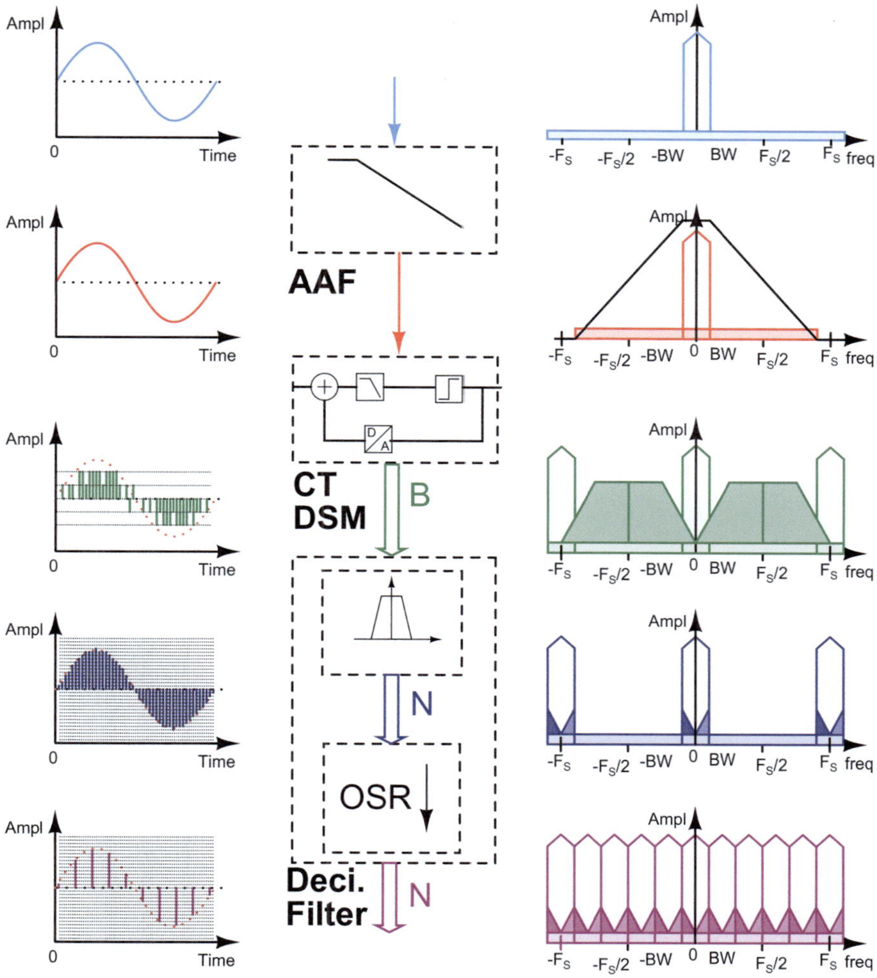

Fig. 3.1 Structure of a whole CT DSM ADC. The corresponding illustrative signals and their spectral plots illustrate the structure's operation

With a much larger sampling frequency f_s than the in-band signal frequency, the STF is close to 1. If the quantizer contains B bits, and the quantization level is Δ, then the output signal power is:

$$P = \frac{1}{2}V_o^2 = \frac{1}{2}(\frac{(2^B - 1)\Delta}{2})^2 \approx \frac{2^{2B}\Delta^2}{8} \tag{3.3}$$

Considering the oversampling and noise-shaping effects, the integrated output quantization noise can be calculated as:

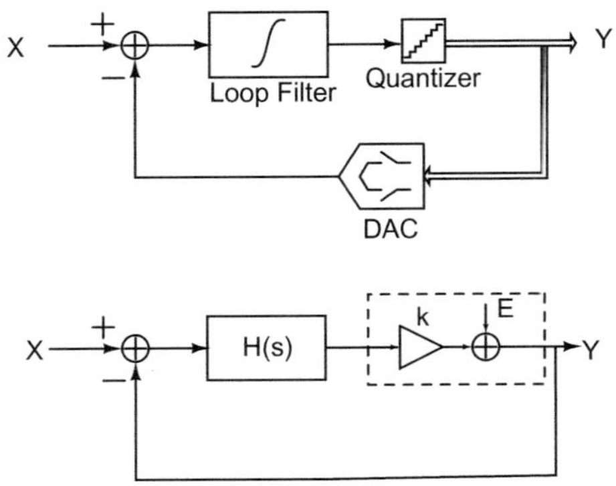

Fig. 3.2 A generalized DSM structure (*top*) and its linear model (*bottom*) for performance analysis

$$
\begin{aligned}
E^2_{q,dsm} &= \int_{-BW}^{BW} |NTF(f)|^2 S_q^2(f) \cdot df \\
&= \int_{-BW}^{BW} \left| \frac{j2\pi f}{f_s} \right|^{2N} \frac{\Delta^2}{12 f_s} df \\
&= \frac{\pi^{2N}}{2N+1} \frac{1}{OSR^{2N+1}} \frac{\Delta^2}{12}
\end{aligned}
\tag{3.4}
$$

So the peak SNR is:

$$
\begin{aligned}
SNR_{p,dsm} &= \frac{P}{E^2_{q,dsm}} = 2^{2B}(2N+1)\frac{3\pi}{2}\left(\frac{OSR}{\pi}\right)^{2N+1} \\
&= 6.02B + 1.76 + 10log\left(\frac{2N+1}{\pi^{2N}}\right) + 10(2N+1)logOSR \ [dB]
\end{aligned}
\tag{3.5}
$$

It is clearly shown in the Eq. 3.5 that the performance of a DSM depends on the number of quantization bits B, the noise-shaping order N and the oversampling ratio OSR. In theory, if the number of quantization bits is increased by 1 bit, then the DSM peak SNR is improved by 6dB. Considering the complexity of the digital part, generally the number of quantization bits is not larger than 5bit. With a doubled OSR, the DSM SNR is improved by $3(2N + 1)$dB. However, the sampling frequency is limited by the maximum speed of each block. For the noise-shaping order N, 5th or higher values would impair the modulator's stability, so MASH structures are proposed. With all these limitations and design considerations, different combinations can be generated, resulting in different DSM structures, which will be presented and discussed in the next section.

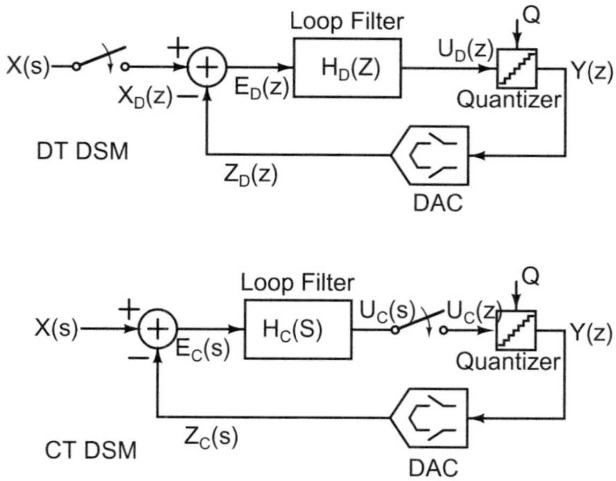

Fig. 3.3 Structure of a DT DSM (*top*) and its CT counterpart (*bottom*)

3.3 DSM Structures

3.3.1 Discrete-Time and Continuous-Time DSMs

DSMs are classified into two basic types: discrete-time (DT) DSMs and continuous-time (CT) DSMs. Their basic structures are shown in Fig. 3.3. Here two differences can easily be observed.

- Different sampling positions: in the DT DSM, the sampling is preceding the whole DSM feedback loop. While in the CT counterpart, the sampling is behind the loop filter, which is inside the feedback loop.
- The loop filter and DAC implementations are different. Loop filters in DT DSMs are based on switched-capacitor (SC) integrators, and CT DSM filters are constructed by active-RC or Gm-C integrators. Commonly used DT and CT integrators are presented in Fig. 3.4. DACs in DT DSMs are also SC circuits, while there are more choices for CT DSM DACs, including current-steering (NRZ and RZ) DACs, SC DACs and switched-resistor DACs, as shown in Fig. 3.5.

Due to their different structures, DT and CT DSMs have different design challenges, bandwidths and performances [3]. The most critical block in a DT DSM is the front-end S/H, as its nonidealities are not shaped by any filter. As the hold capacitor value is determined by the kT/C noise, a small value and a high linearity are required from the switch on-resistance. Since bipolar technology is not a good choice for such a switch design, DT DSMs became popular only after the introduction of CMOS technology. With CMOS technology scaling, the transmission gate on-resistance becoming larger and larger, it is more and more difficult to design high-accuracy DT DSMs with simple switches. To tackle this challenge, both bootstrapping switches [4] and switched-OpAmp [5] techniques have been proposed. In a

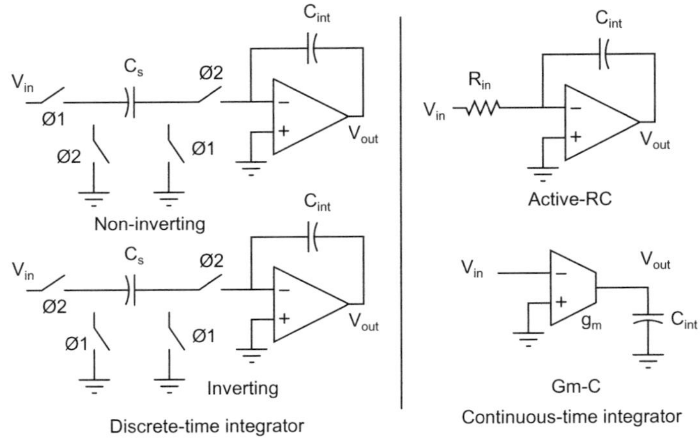

Fig. 3.4 Popular circuit implementations of DT (*left*) and CT (*right*) integrators used in the DSM loop filter

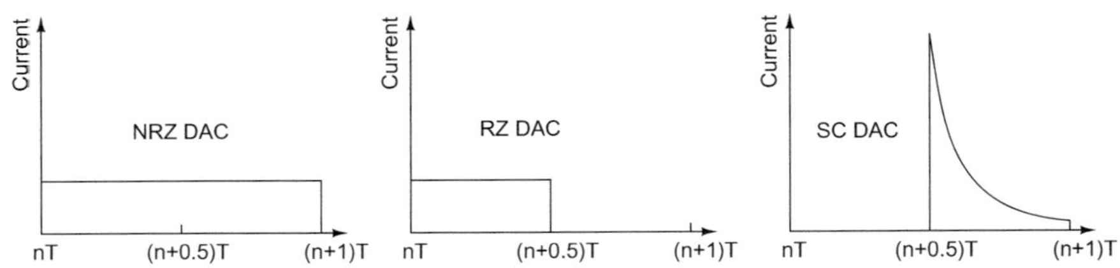

Fig. 3.5 Different DAC current profiles used in CT DSMs: NRZ (*left*), RZ (*middle*) and SC (*right*)

CT DSM, being shaped by the preceding analog loop filter, the quantizer switch nonidealities almost do not contribute to the modulator performance degradation. For a DAC cell, its input has only two states (1 and 0), and there is no switch on-resistance nonlinearity problem.

In DT DSMs, the achievable SNR depends on the SC circuit settling accuracy, which requires OTAs with larger SR and GBW than CT ones. Generally speaking, an OTA GBW of at least $5\times$ the sampling frequency is needed in DT DSMs, while the ratio in CT DSMs is only 1 to $3\times$. With the larger ratio between the OTA GBW and the sampling frequency, the analog power consumption in DT DSMs is higher and the maximum allowed sampling frequency is lower. With existing CMOS technology, DT DSMs are suitable for bandwidths <10MHz and accuracies >12bit designs.

In a DT DSM, the feedback loop can tolerate up to one period delay. As a result, any delay in the CT DSM loop will affect its stability and changes the NTF. To accommodate the excess loop delay (ELD) in CT DSMs, half-period and full-period compensations are widely adopted and an extra zero-order feedback path is added to maintain the stability and to keep the NTF the same. Half-period compensation tolerates a smaller ELD, but with a larger delay budget for the DAC and loop filter, relaxing the speed requirement for these blocks.

Once the SC circuit settling in the DT DSM is sufficient, it is very robust to clock jitter. This is different in CT DSMs, where the DAC is more sensitive to clock jitter, especially when a RZ DAC or a single-bit DAC is applied. To address the clock jitter noise issue, several solutions can be adopted in CT DSM design, for example multi-bit qantization together with NRZ DAC [6], SC DAC [7] and FIR single-bit DAC [8] techniques. With such performance limitations, CT DSMs are mostly used for high-bandwidth (>10MHz) moderate-accuracy (10-12bit) applications.

Another disadvantage of CT DSMs compared to their DT counterparts is their large variation in coefficients. In SC integrators, the SC part behaves as an equivalent resistor, and the integrator gain is determined by the ratio between the integration capacitor and the feedback capacitor (if the OTA is ideal):

$$R_{eq} = \frac{1}{f_s C_s} \tag{3.6}$$

$$H(z) = \pm \frac{C_s}{C_{int}} \frac{z^{-0.5}}{1 - z^{-1}} \tag{3.7}$$

With good capacitor matching in CMOS technology, this ratio can reach a very high accuracy. Differently, the CT integrator transfer function is based on the absolute values of the resistor transconductor and capacitor:

$$H(s) = -\frac{1}{sR_{in}C_{int}} \tag{3.8}$$

$$H(s) = -\frac{g_m}{sC_{int}} \tag{3.9}$$

In CMOS design, the absolute accuracies of single component value are not under good control: the variation can be as high as $\pm 20\%$. So in a CT DSM design, a capacitor bank [9] is always required to adjust the integrator gain.

One merit of DSMs compared to Nyquist ADCs is the relaxed preceding AAF. In DT DSMs, because of the front-end sampling and the DT loop filter, the AAF is relaxed by the oversampling. In CT DSMs, since the quantizer is behind the analog loop filter, the AAF is not only relaxed by the oversampling, but also by the intrinsic filtering of the loop filter. If a feedback loop filter topology is selected, usually the separate AAF can be eliminated.

Although there are several differences between the DT and CT DSMs, in the CT DSM design flow, it is usually preferred that first a DT NTF is designed, which is then converted to the desired CT NTF by using the impulse invariant transformation (IIT) [10].

3.3.2 1st-Order and Higher-Order DSMs

In the 1st-order DSM, the loop filter is only an integrator, which introduces a 90° degree phase shift. Considering the 180° degree phase shift caused by the negative feedback, the phase margin (PM) of a 1st-order DSM is 90° degree, which means that it is unconditionally stable if it is not overloaded. However, in the 1st-order DSM, the quantizer input is not random and its quantization error can not be treated as white noise. The consequence is that the 1st-order DSM is susceptible to limit cycles, which means that its output contains some pattern noise if its analog input is a DC signal. One proof of the limit cycles is that the SNR of the 1st-order DSM with DC input highly depends on the input amplitude. To solve this limit cycle problem, a white noise used for dithering can be injected into the DSM loop [11]. Another drawback of a 1st-order DSM is the much lower performance compared with higher-order DSMs. In Fig. 3.6, the SNR of a Nth-order ($N = 1 \cdots 5$) 3-bit DSM with different OSRs is plotted using Eq. (3.5). It can be concluded that a much larger OSR is needed for a 1st-order DSM to achieve the same performance as higher-order modulators.

In some high-performance DSM design, the required OSR for 1st-order DSM may be too high to be implemented in the technology. One solution to the too high sampling frequency is to adopt a higher-order DSM. In a 2nd-order DSM, there is more phase shift in the loop filter and it is only conditionally stable. If the gain of the first integrator is larger than 1, there are both stable and unstable limit cycles, resulting into a unstable modulator. To design a stable 2nd-order DSM, the first integrator gain should be less than 1 and the modulator can not be overloaded. 3rd-order and higher-order DSMs are even more prone to instability, including DC instability and dynamic instability. According to Lee's empirical rule [12], a single-bit DSM is stable if its maximum out-of-band gain H_{inf} is not larger than 2. However, this criterion is proved not necessary nor sufficient. To design a stable higher-order DSM, several solutions are helpful.

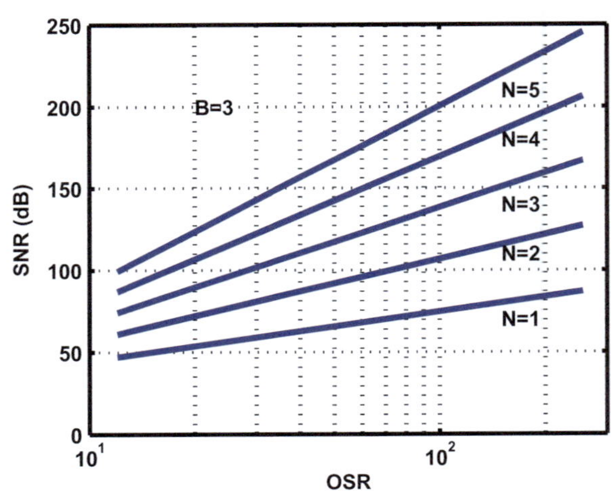

Fig. 3.6 Performance of Nth-order 3-bit DSM with different OSRs

1. Use the proven design tool to synthesize the NTF and loop filter, for example the Delta-Sigma toolbox [1] based on Matlab, which is developed by Richard Schreier.
2. Scale the integrator gain and the input signal swing, which will however degrade the DSM performance.
3. A RC filter preceding a high-order DSM can effectively reduce the dynamic instability problem.
4. Use a clamp circuit to limit the integrator voltage swing to the normal stable operating range.
5. Reset the integrator capacitor once the integrator output is beyond the stable range or there are long strings of consecutive 1's and 0's at the output.
6. Use multi-bit quantization and/or MASH structures, which will however introduce other design issues, as presented later.

3.3.3 Single-Loop and MASH DSMs

As discussed in the last subsection, higher-order DSMs suffer from stability issues and consequently have a smaller overload level. For DSMs with 3rd- or higher-order noise shapings, to keep the same noise-shaping order and simultaneously avoid the potential instability problem, the multi-stage noise-shaping (MASH) topology [13] has been proposed. A DT 2-2 MASH DSM is shown in Fig. 3.7. In this structure, there are two stages and each stage is a single-loop 2nd-order modulator. The quantization noise of the first stage E_1 is obtained by a subtractor and used as the input of the second stage. The outputs of both stages are respectively:

$$Y_1(z) = X(z)z^{-2} + E_1(z)(1 - z^{-1})^2 \tag{3.10}$$

$$Y_2(z) = -E_1(z)z^{-2} + E_2(z)(1 - z^{-1})^2 \tag{3.11}$$

The digital outputs of the two stages are processed further by the digital noise cancellation filter (DNCF) and its output is the output of the entire MASH DSM:

$$\begin{aligned} Y(z) &= Y_1' + Y_2' = Y_1 z^{-2} + Y_2(1 - z^{-1})^2 \\ &= X(z)z^{-4} + E_1(z)(1 - z^{-1})^2 z^{-2} \\ &- E_1(z)z^{-2}(1 - z^{-1})^2 + E_2(z)(1 - z^{-1})^4 \\ &= X(z)z^{-4} + E_2(z)(1 - z^{-1})^4 \end{aligned} \tag{3.12}$$

It is shown in this equation that if the DNCF is properly designed, the quantization noise of the first-stage DSM E_1 is totally cancelled and the quantization noise of the second stage E_2 is 4th-order noise-shaped. In a real implementation, the noise cancellation depends on the matching between the NTF of the first-stage DSM and

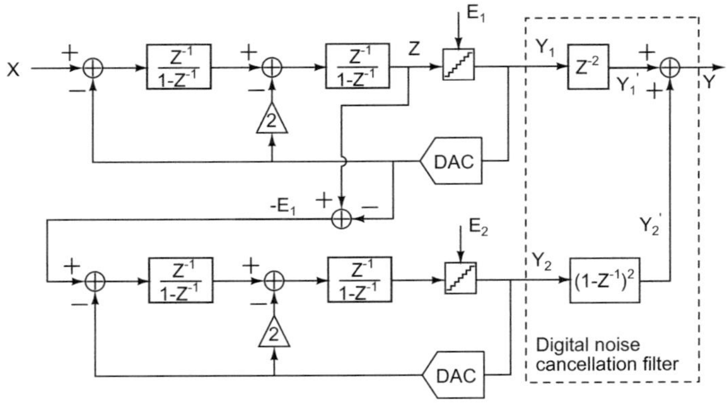

Fig. 3.7 The architecture of a discrete-time 2-2 MASH DSM, the total noise shaping is 4th-order

the DNCF $(1-z^{-1})^2$. If the matching is not perfect, part of the first-stage quantization noise E_1, which is only 2nd-order noise-shaped, appears at the DSM output. This is called MASH DSM noise leakage. In a DT modulator realization, although both the NTF and the DNCF are implemented in DT, lots of circuit nonidealities affect the NTF, including the OTA limited slew-rate, the GBW, etc. So the requirements for the building blocks in the MASH structure are more rigorous than in a single-loop one.

In CT MASH DSMs, the NTF is analog while the DNCF remains digital. These different natures make the function matching very difficult. Furthermore, compared to the DT NTF, the CT NTF is more sensitive to circuit nonidealities, which will be covered in the next section. As a result, it is much more difficult to achieve a good function matching and noise cancellation. This is the reason why most of the MASH DSMs are realized in DT. To relax the matching requirement and to ease the block design, usually digital calibration is applied in CT MASH structures [14] to tune the path gain in the DNCF. In CT MASH DSMs, all stages not only contribute to the total noise shaping, but also provide AAF for the input signal [14]. Since all noise and nonlinearity of the second stage is shaped by the first stage, the design of the following stages can be relaxed greatly to reduce the overall power consumption and area.

3.3.4 The $\Delta\Sigma$-0 and 0-$\Delta\Sigma$ MASH Structures

Specially, an open-loop quantizer without preceding filters, or so-called a zero-order quantizer, can be adopted in the stages of the MASH modulator. With a zero-order stage, a two-stage MASH $\Delta\Sigma$ ADC can be implemented with a $\Delta\Sigma$-0 or a 0-$\Delta\Sigma$ structure. This subsection will discuss about the performance of these structures.

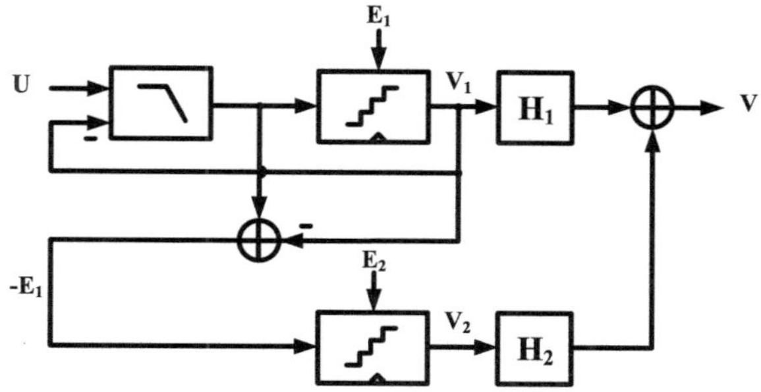

Fig. 3.8 Block diagram of a $\Delta\Sigma$-0 MASH structure

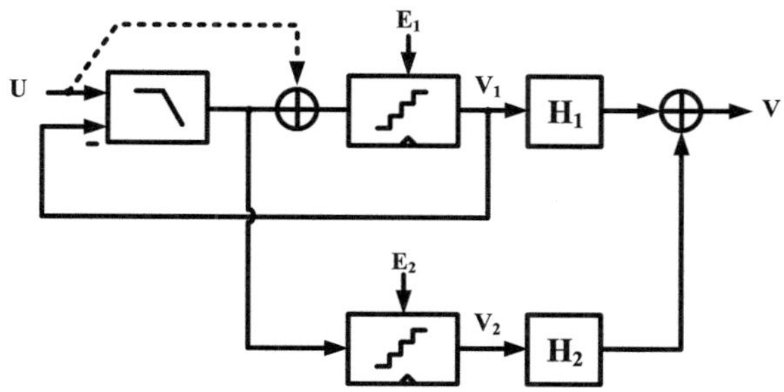

Fig. 3.9 Block diagram of a modified $\Delta\Sigma$-0 MASH structure

As shown in Fig. 3.8, if the zero-order quantizer is used as the second stage, the cascade structure is a $\Delta\Sigma$-0 MASH, also known as the Leslie-Singh architecture [15]. Here, $H_1 = z^{-k}$ is a delay implemented to compensate the latency of the 2nd stage, and $H_2 = NTF_1$. Then the whole modulator output becomes:

$$V = H_1 \cdot STF_1 \cdot U - H_2 \cdot NTF_2 \cdot E_2 = z^{-k} \cdot (STF_1 \cdot U - NTF_1 \cdot E_2) \tag{3.13}$$

Compared to a single-loop modulator, E_2 replaces E_1 appearing at the output. Since the 2nd stage is an open-loop structure, it has a relaxed latency requirement, and thus can be realized with a low-complexity multi-bit pipeline structure [1], making E_2 much smaller than E_1. Therefore, the peak SQNR performance can be enhanced. However, the matching requirement $H_2 = NTF_1$ is still difficult to be satisfied.

A modified $\Delta\Sigma$-0 MASH structure is shown in Fig. 3.9 to eliminate the substractor preceeding the 2nd stage. Keeping $H_1 = z^{-k}$ and choosing $H_2 = \mathrm{NTF}_1/(\mathrm{NTF}_1 - 1)$, the final output becomes:

$$V = \frac{z^{-k} \cdot STF_1}{1 - NTF_1} \cdot U + \frac{z^{-k} \cdot NTF_1}{1 - NTF_1} \cdot E_2 \tag{3.14}$$

In the signal band, it is safe to assume $|NTF| \ll 1$, and therefore the result is close to Eq. (3.13). However, the drawback of this structure is that the input of the 2nd stage may contain the input U. To aviod that the 2nd stage needs to handle a

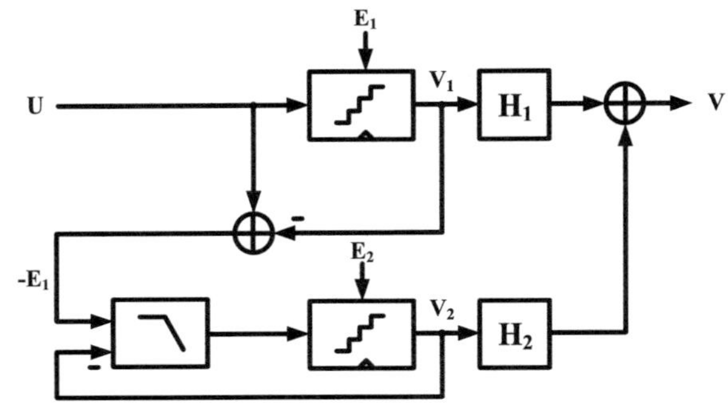

Fig. 3.10 Block diagram of a 0-$\Delta\Sigma$ MASH structure

large input signal, a feedforward topology should be implemented for the integrators in the $\Delta\Sigma$ loop of the first stage.

Another structure is the 0-$\Delta\Sigma$ MASH, as presented in Fig. 3.10. A crucial insight from the above discussion is that, if the first stage is a zero-order open-loop quantizer with an approximately unity NTF_1, resulting in the constraint $H_2 = NTF_1 = 1$, then the matching requirement can be relaxed. Although the 2nd stage has the capability of processing signals above the full scale (FS) of the 1st-stage quantizer [16], the input signal is typically limited within the FS range. Therefore, the 2nd stage only processes the quantization noise, and power-efficient opamps with improved linearity can be implemented here. The final output of the 0-$\Delta\Sigma$ MASH becomes:

$$V = H_1 \cdot U + (H_1 - STF_2) \cdot E_1 - NTF_2 \cdot E_2 \tag{3.15}$$

However, different from Eq. (3.12), the quantization noise E_2 is only shaped by NTF_2, which is the same as is the case in a single-loop structure. Unfortunately, the first stage consumes power, but cannot offer additional noise shaping.

3.3.5 *Single-Bit and Multi-bit DSMs*

In single-bit DSMs, the quantizer is only one comparator. With its small input parasitic capacitor, the comparator can achieve a very high speed. This explains why most of the high-sampling-frequency DSMs [17, 18] choose the single-bit topology. In single-bit DSMs, compared to the analog part, the power consumption and chip area of the digital part is only a small fraction. On the other hand, the comparator output only depends on the polarity of its input, but not on the input signal amplitude, as shown in the right part of Fig. 3.11. This means that the gain of a single-bit quantizer is highly nonlinear, as illustrated in the left part of Fig. 3.11. Because of this nonlinearity, the single-bit DSM is more prone to instability than multi-bit DSMs and has a lower overload level.

Correspondingly, the DAC of a single-bit DSM also is the simplest case, as it has only two output levels. With only two levels, no matter what values they have, they

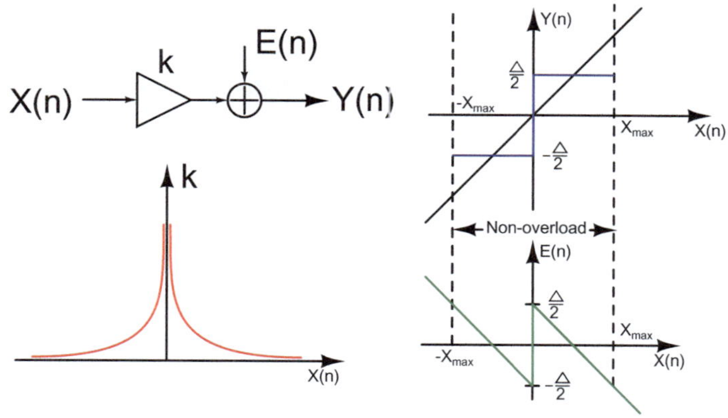

Fig. 3.11 Single-bit quantizer: the linear model, the input-output transfer function and its highly nonlinear gain

can be connected by a straight line. In other words, a single-bit DAC is intrinsically linear, which is its most important advantage. The intrinsic linearity makes single-bit DSMs very suitable for high-linearity designs, e.g. digital audio. However, if a NRZ CS DAC is applied, a single-bit DSM is more sensitive to clock jitter than multi-bit DSMs, as the clock-jitter-induced noise is proportional to the DAC output step variation. One solution to the single-bit DSM clock jitter issue is to insert a digital FIR multi-tap filter between the quantizer and the DAC, in such a way that the DAC behaves as a multi-bit DAC [8]. Another challenge of the single-bit DSM is that there is more quantization noise generated than in multi-bit DSMs, and the amplitude of the loop filter residual signal is larger, so a larger voltage swing and gain are needed in single-bit DSMs.

If a multi-bit quantization is applied in a DSM design, different advantages and disadvantages are obtained. First, with more quantization steps, less quantization error is introduced to the feedback loop. Every extra bit of quantization results in DSM SNR improvement of 6dB. Using a multi-bit topology, the feedback DAC output tracks the analog input signal more accurately. This leads to a smaller residual signal in the loop filter, relaxing the filter amplifier design. Another positive aspect of multi-bit DSMs is that the quantizer gain is well defined and so the modulator stability is better under control. Similar to a flash ADC design, the multi-bit quantizer can not operate as fast as a simple comparator: the input parasitic capacitor together with the output impedance of the proceeding stage introduces another pole and limits the sampling frequency. The exponentially increasing hardware of a multi-bit quantizer also costs considerable power and chip area.

For the DAC of a multi-bit DSM, especially the front-end DAC, its performance tradeoff is very meaningful for the whole DSM design, as its nonidealities determine the DSM SNDR and SFDR directly. First of all, a multi-bit NRZ DAC is more robust to clock jitter than a single-bit one. In an implementation, because of the component's systematic and random variations, the values of different cells in a multi-bit DAC are

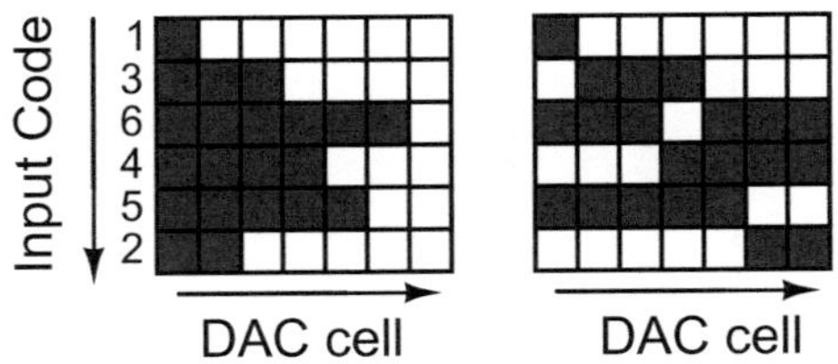

Fig. 3.12 The concept of DEM: 3-bit DACs without (*left*) and with (*right*) DEM are illustrated for comparison

different, which is called multi-bit DAC mismatch. One serious result of the DAC mismatch is the appearance of harmonic distortion at the DSM digital output. To solve the multi-bit DAC mismatch and to improve the DSM linearity, three kinds of techniques can be applied:

1. Design the DAC current source array with sufficient area and a proper layout (e.g. the centroid, double centroid, triple centroid) to reduce the static mismatch. The driving circuit for each DAC cell should also be designed carefully to eliminate any dynamic mismatch. A multi-bit DSM with 12-bit intrinsic DAC matching accuracy has been reported [9] in this way.
2. Use a dynamic element matching (DEM) technique [19], the concept of which is shown in Fig. 3.12. In a DAC without DEM, the first DAC cell is used most frequently, and the probabilities of cells being under used decrease with their number. If a DEM block is placed preceding the DAC, then all cells are used almost equally. By doing so, the DAC mismatch is 1st-order shaped. However, a high-bit DEM generates a large delay, which may destroy the modulator stability, thus it is not a good choice for high-bit DSMs with large sampling frequencies. The high-bit DEM block will also consume considerable digital power and area.
3. Calibrate the DAC mismatch via a digital look-up table (LUT) [20]. In such a calibration, in the extraction phase first the DAC cell mismatch information is extracted and stored in a LUT. Then in the use phase, based on the quantizer output, a calibrated value is withdrawn from the table as the final DSM output. The benefit of digital LUT-based calibration is that it is out of the feedback loop and its delay does not affect the DSM stability. The drawback of digital calibration is that it can not track different variations, including temperature and aging variations.

3.3.6 Feedforward, Feedback and Hybrid DSMs

Once the DSM NTF is synthesized, the loop filter transfer function can be deduced. Several kinds of topologies are available for the filter realization and they all shown their merits and drawbacks. In Fig. 3.13, generalized Nth-order DSMs with four

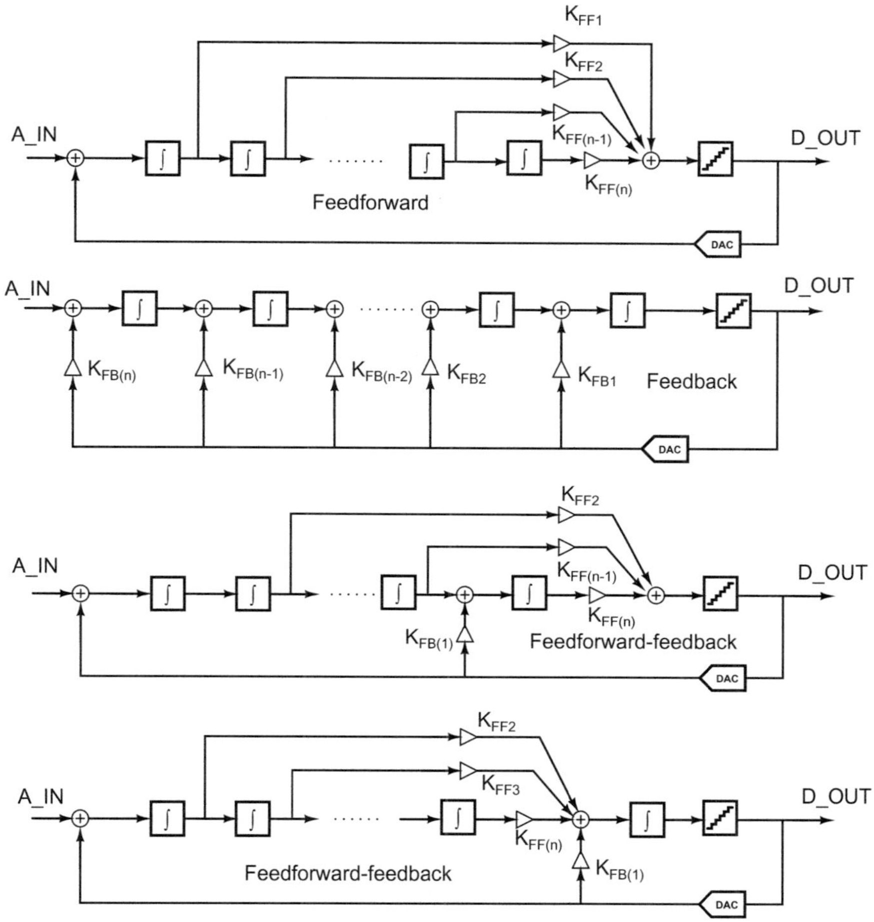

Fig. 3.13 Different structures of generalized Nth-order DSMs: feedforward (*top*), feedback (*upper middle*) and hybrid ones (*lower middle* and *bottom*)

different kinds of filter topologies (one feedforward, one feedback and two hybrid ones) are illustrated. They have the same NTF but different STFs.

The first architecture is the feedforward DSM [21]. Here, there is only one (front-end) feedback DAC. The outputs of all integrators are summed with different weights and fed to the quantizer. The benefit of the feedforward DSM is that only noise goes through the loop filter, so the output swings of the integrator OTAs are relaxed. However, one extra adder is needed preceding the quantizer. An active adder consumes extra power and a passive one can only provide attenuation. For the STF of a feedforward DSM, its AAF has only a 1st-order roll-off, which limits its application potential.

Another widely used topology is the feedback one [22], where in total N DACs are needed to feed the output signal with different gains back to the loop filter, which is its biggest shortcoming. The output of the *i*th integrator minus the (i+1)th DAC output results in the residual signal for the (i+1)th integrator. One important advantage of the feedback-type DSM is that no extra adder is needed. Moreover, a feedback DSM has a smaller STF in-band ripple than a feedforward DSM. Last but not least, the AAF characteristic of the feedback DSM STF has a Nth-order roll-off, which can greatly relax or even eliminate the preceding AAF.

To make use of the merits of both the feedforward and feedback topologies, hybrid DSM structures have been proposed. The first hybrid architecture is similar to the feedforward one, except that the 1st-order feedback loop is composed of the last integrator, but not the first integrator. To do that, one more DAC is needed to feed back the signal into the last integrator. In such topology, the extra adder is still needed, but with one input less.

To eliminate the extra adder in the hybrid structure, one popular solution is to combine the last integrator and the adder together. In such a case, the 2nd-order feedback loop is not constructed by the first two integrators, but by the first and the last integrators. In both hybrid topologies, the STF AAFs have a 2nd-order roll-off. With different STF AAF orders and different DAC numbers, several different hybrid topologies can be applied in design, depending on the exact design specifications and considerations.

3.3.7 Resonator

With all the kinds of loop filter topologies as seen in the above subsection, all poles of the loop filter are at the DC frequency, and so are the NTF zeroes. In 3rd- and higher-order DSMs, if some of NTF zeroes are moved from the DC frequency to higher non-zero frequencies, the integrated in-band quantization error can be several dB lower. This SQNR improvement due to a NTF zero distribution optimization is more obvious in high-bandwidth DSMs. In the circuit implementation, such complex non-zero NTF zero pair is generated by a resonator in the loop filter, as presented in Fig. 3.14. The resonator is a negative feedback loop across two integrators, with a 180° phase shift from the negative feedback and a 90° phase shift from each integrator. The resonator in Fig. 3.14 introduces two NTF zeroes at the frequency of:

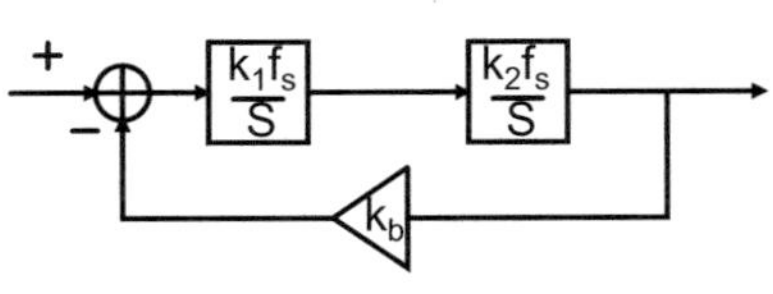

Fig. 3.14 Resonator in the loop filter for the generation of complex NTF zeroes

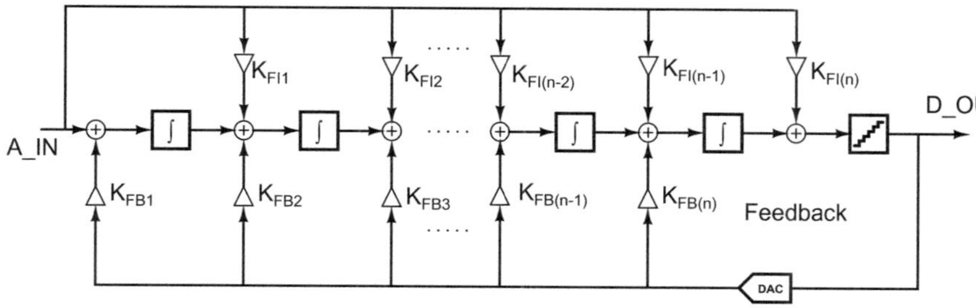

Fig. 3.15 Direct coupling paths from the DSM input to integrators and quantizer for integrator output swing reduction

$$f = \frac{\sqrt{k_b k_1 k_2 f_s}}{2\pi} \tag{3.16}$$

3.3.8 *Feedin Paths*

As shown in Sect. 3.3.6, DSMs especially with a feedback loop filter face the design challenge of having large OTA output swings. To resolve the stringent OTA swing requirement and to reach high linearity, direct coupling paths from the modulator analog input to the inputs of integrators and the quantizer are added in a feedback-type DSM, as shown in Fig. 3.15. If the gains of these coupling paths are designed properly, the direct-coupled signals cancel the DAC outputs and only noise is processed by the integrators, relaxing the OTA output swings largely. These direct-coupling paths can also be applied in feedforward-type DSMs [21]. However, the integrator output swing reduction and linearity improvement is not so obvious. In a real circuit implementation, not all direct-coupling paths are necessary. For integrators closer to the quantizer, their nonlinearities are suppressed by the preceding stages, so the corresponding direct coupling-paths can be removed.

These direct-coupling paths do not change the modulator NTF, but they modify the STF in two aspects. First, with these direct-coupling paths, the in-band ripple of the STF becomes worse. Furthermore, the STF AAF characteristic degrades in a CT DSM, especially in a feedback-type DSM. With such paths, the AAF no longer has a Nth-order roll-off, but a kth-order roll-off ($0 \leq k \leq (N-1)$), depending on the injection positions of the paths.

3.4 CT DSM Nonidealities and Modeling

In CMOS circuit implementations, no building block of a CT DSM can be perfect. Different circuit nonidealities have different mechanisms and different effects on the modulator performance [23]. The nonideality effect also highly depends on the location. Generally speaking, the front-end nonidealities are more critical than the back-end ones, as the latter ones are suppressed by the preceding gain stage.

3.4.1 Loop Filter Nonidealities and Modeling

The loop filter is important for a DSM design, since the modulator's noise shaping capability is determined by the loop filter transfer function. Inside the loop filter, the first stage is the most critical one, as its nonidealities are not suppressed by any gain, determining the DSM performance directly. Both active-RC and Gm-C integrators can be used in the loop filter design, and the active-RC one is typically applied in high-accuracy (>10-bit) DSMs.

3.4.1.1 Finite OTA Gain

In a CT DSM design, without stringent settling requirement, the OTA requirement is relaxed compared to a DT DSM. However, amplifier nonidealities still introduce lots of errors, degrading the modulator performance or even destroying its stability. For signal summation and resonator generation, an active-RC integrator in a CT DSM can have as many as three inputs, as shown in Fig. 3.16. If the OTA is ideal, the transfer function from the ith input to the output of the active-RC integrator is:

$$H(s)_i = -\frac{1}{sR_{in(i)}C_{int}} = -\frac{k_i f_s}{s} \tag{3.17}$$

Its pole is located at DC, thus it has an infinite gain at DC. First assume that the OTA gain is A_{DC} at any frequency:

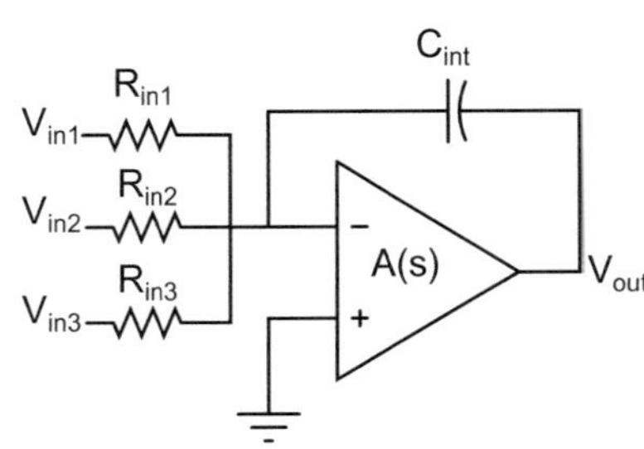

Fig. 3.16 An active-RC integrator with nonideal OTA and three input voltages in a CT DSM

$$A(s) = A_{DC} \tag{3.18}$$

Then the integrator transfer function can be derived as:

$$\begin{aligned} H(s)_i &= -\frac{1}{R_{in(i)}C_{int}} \frac{A_{DC}}{s\,(A_{DC}+1) + \sum_{i=1}^{3} \frac{1}{R_{in(i)}C_{int}}} \\ &= \frac{k_i f_s}{s\left(1+\frac{1}{A_{DC}}\right) + \frac{\sum_{i=1}^{3} k_i f_s}{A_{DC}}} \end{aligned} \tag{3.19}$$

Hence the integrator pole is shifted from DC to $\sum_{i=1}^{3} k_i f_s/(1+A_{DC})$ and its gain (especially at low frequency) is reduced. Since the loop filter in-band gain determines the suppression for the quantization error and the other circuit nonidealities, the limited OTA gain results in a noise leakage phenomenon. If a certain integrator gain is required at the signal bandwidth, it can be concluded that the OTA gain requirement relies on the OSR and the integrator coefficient. A larger OSR always calls for a larger amplifier gain. The relationship between the required amplifier gain and the integrator coefficients is however more complex and can be investigated by behavioural simulations. A rule of thumb for the required OTA gain in a single-loop DSM is that it should not be less than the OSR. Another effect of the OTA finite gain is that the signal swing of the OTA input nodes is not zero anymore. For a differential OTA input pair, the linear input range is around [-$\sqrt{2}(V_{gs}-V_{th})$, $\sqrt{2}(V_{gs}-V_{th})$], here the V_{gs} and V_{th} are the transistor gate-source and threshold voltages respectively. The too large input voltage swing together with the nonlinear OTA input stage introduce harmonic distortion in the CT DSM output PSD.

3.4.1.2 Finite OTA GBW

Next, we consider the OTA in an active-RC integrator as a 1st-order block, with not only a finite gain, but also a finite bandwidth:

$$A(s) = \frac{A_{DC}}{1+\frac{s}{\omega_p}} = \frac{1}{\frac{1}{A_{DC}} + \frac{s}{GBW}} \tag{3.20}$$

Applying this expression to the equation of the integrator, we obtain the nonideal transfer function:

$$
\begin{aligned}
H(s)_i &= \\
&-\frac{1}{R_{in(i)}C_{int}}\frac{1}{\frac{1}{GBW}s^2+\left(1+\frac{1}{A_{DC}}+\frac{1}{GBW}\sum_{i=1}^{3}\frac{1}{R_{in(i)}C_{int}}\right)s+\frac{1}{A_{DC}}\sum_{i=1}^{3}\frac{1}{R_{in(i)}C_{int}}} \\
&= -\frac{1}{R_{in(i)}C_{int}}\frac{1}{\frac{1}{GBW}s^2+\left(1+\frac{1}{GBW}\sum_{i=1}^{3}\frac{1}{R_{in(i)}C_{int}}\right)s} \\
&= -\frac{1}{sR_{in(i)}C_{int}}\frac{1-\frac{\sum_{i=1}^{3}\frac{1}{R_{in(i)}C_{int}}}{GBW+\sum_{i=1}^{3}\frac{1}{R_{in(i)}C_{int}}}}{1+\frac{s}{GBW+\sum_{i=1}^{3}\frac{1}{R_{in(i)}C_{int}}}} \\
&= -\frac{k_if_s}{s}\frac{1-G_E}{1+\frac{s}{\omega_E}}
\end{aligned} \tag{3.21}
$$

where G_E and ω_E stand for the gain error and the phase error respectively:

$$
G_E = \frac{\sum_{i=1}^{3} k_if_s}{GBW + \sum_{i=1}^{3} k_if_s} \tag{3.22}
$$

$$
\omega_E = GBW + \sum_{i=1}^{3} k_if_s \tag{3.23}
$$

From the above derivation, the finite OTA GBW introduces two extra terms to the active-RC integrator transfer function: a gain error and an extra pole. The gain error reduces the loop filter suppression capability for quantization noise and other nonidealities. The more severe consequence is that the extra pole generates extra phase shift and time delay for the DAC signal (high-frequency signal) when it passes through the integrator, changing the modulator NTF and degrading the modulator stability. It is clear that a large OTA GBW is helpful for building a high-performance active-RC integrator, at the cost of extra power consumption. If the OTA GBW is fixed, for the integrator coefficient value, there is a trade-off between the gain and phase errors. In our designs later on, the integrator gain is decided by parameter explorations in behavioural simulations.

3.4.1.3 Parasitic Capacitors of the OTA and the Integration Capacitor

Another nonideality of an active-RC integrator is the input parasitic capacitance, including the OTA input parasitic capacitance and the parasitic capacitance of the integration capacitor (positive end). To investigate its effect on the integrator performance, a single-input active-RC integrator with input parasitic capacitor C_{in} is presented in Fig. 3.17. Here the OTA transfer function is the same as in the above analysis. Now the transfer function of the nonideal active-RC integrator is:

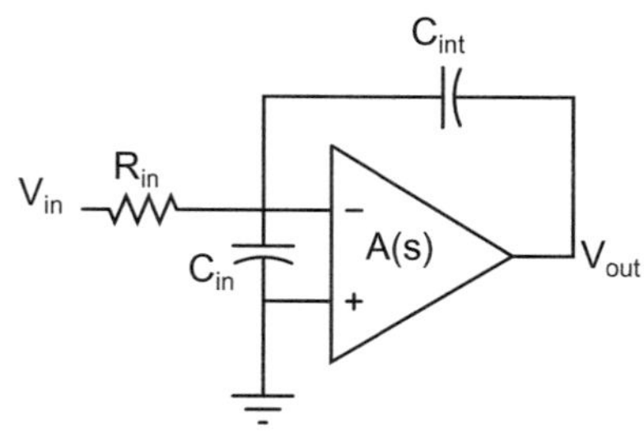

Fig. 3.17 An active-RC integrator with nonideal OTA and input parasitic capacitor

$$H(s) = -\frac{kf_s}{s}\frac{GBW}{GBW + (\omega_p + s)\left(\frac{C_{in}}{C_{int}} + 1 + \frac{kf_s}{s}\right)} \tag{3.24}$$

As expected, the OTA input parasitic capacitance C_{in} degrades the integrator performance further, and the extra degradation depends on the OTA GBW and the ratio C_{in}/C_{int}. A rule of thumb is to keep the input parasitic capacitance less than one third of the integration capacitor. The output parasitic capacitance, including the OTA output parasitic capacitance and the parasitic capacitance of the integration capacitor (negative end) does not change the integrator transfer function. However, extra power consumption is needed in the OTA to drive these capacitances.

3.4.1.4 Limited OTA Output Swing

The final important OTA nonideality is its limited output swing. If the integrator output swing is larger than the OTA output swing, the integrator output is clamped and distortions are generated. This nonideality is more severe when nanoscale CMOS technology is used in the design, as the available headroom voltage is much smaller. It is essential to note that the effects of these mentioned OTA nonidealities depend on the system-level parameters, i.e. the OSR, the loop filter architecture and the number of quantization bits.

In behavioral-level modeling, one method is to deduce the transfer function of the whole integrator with a nonideal OTA, and symbolically express the error induced by the OTA nonidealities. Another solution is to use the model from Fig. 3.18. Here the complex nonideal integrator transfer function is modeled and more OTA nonidealities can be added, resulting in a higher-accuracy model. The OTA input stage nonlinearity and the output limited swing are represented by a hyperbolic tangent function and a signal clamp, respectively. The limited gain, GBW and doublet (comes from feedforward frequency compensation) are represented by the linear filter stages in Matlab Simulink. In this model, the effect of the input parasitic capacitor can also be considered conveniently, by calculating the transfer function of the two voltage dividers with one extra component.

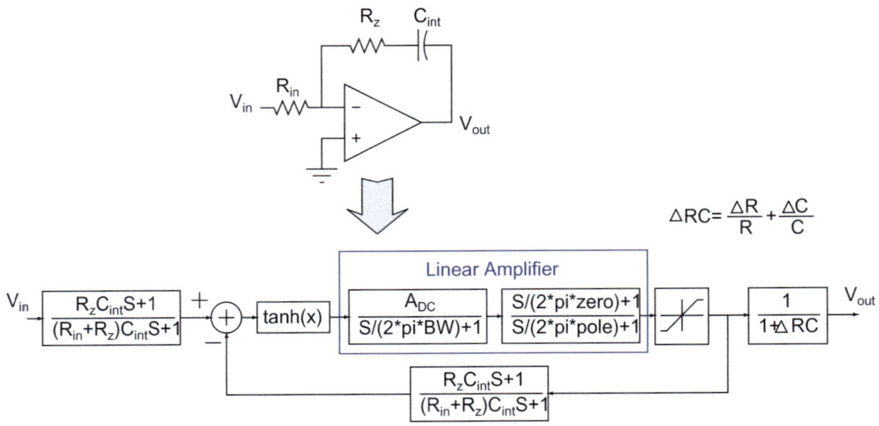

Fig. 3.18 An accurate and convenient nonideal model of an active-RC integrator in Matlab Simulink

3.4.1.5 RC Time Constant Variation

Besides the nonideal OTA, the input resistor and the integration capacitor also effect the CT DSM performance, as the gain of an ideal active-RC integrator is determined by their absolute values. In CMOS technology, the variation on passive component absolute value can be as high as $\pm 30\%$. As a result, the active-RC integrator gain deviation can be as large as $[-41\%, +104\%]$. With a negative gain variation, the suppression of the quantization error and other circuit nonidealities will not be so effective and the CT DSM performance is degraded. On the other hand, when the integrator gain varies positively, the shaping of noise and other nonidealities is better, increasing the modulator SNDR. However, once the integrator gain exceeds some threshold level, the CT DSM becomes unstable. In behavioral-level simulations, a gain error stage can be added to model the RC time constant variation (see Fig. 3.18):

$$H'(s) = -\frac{1}{S(R+\Delta R)(C+\Delta C)} \approx -\frac{1}{SRC}\frac{1}{1+\Delta R/R+\Delta C/C} \tag{3.25}$$

To design a CT DSM with stable performance, its RC time constant needs to be tuned, which is done by a variable capacitor. In the variable capacitor in Fig. 3.19, a capacitor C_o with fixed value is connected in parallel with a 3-bit binary-controlled capacitor bank. The normal, minimum and maximum capacitor values are $C_o + 3C_u$, C_o and $C_o + 7C_u$ respectively. Thus the reachable tuning range and step are $[-\frac{3C_u}{C_o+3C_u}, \frac{4C_u}{C_o+3C_u}]$ and $\frac{C_u}{C_o+3C_u}$ respectively. To reduce the nonlinearity, both the zero-cancellation resistor R_o and the switches are connected to the integrator vitual ground, whose signal swing is smaller than that of the integrator output.

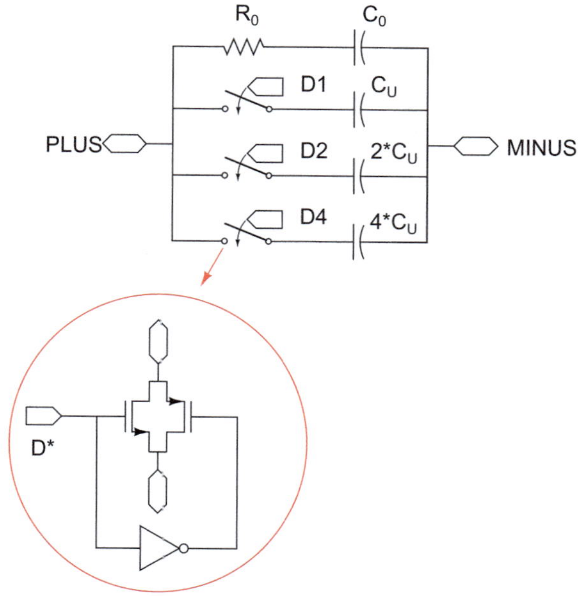

Fig. 3.19 Variable capacitor for RC time constant tuning in an active-RC integrator

3.4.2 *DAC Nonidealities and Modelling*

In a CT DSM, the front-end DAC directly influences the modulator performance. Thus the DAC nonidealities must be modelled and addressed in the different design phases, including clock-jitter-induced noise, inter-symbol interference (ISI) and multi-bit DAC mismatch.

3.4.2.1 Clock-Jitter-Induced Noise

Clock jitter is the manifestation in the time domain of phase noise. It can be classified into two types: the pulse-delay jitter and the pulse-width jitter. The effect of the pulse-delay jitter on the CT DSM performance can be neglected, as it can be considered as a variable loop delay and such delay can be tolerated by the modulator. On the other hand, the pulse-width jitter stochastically changes the amount of charge transferred from the feedback DAC to the loop filter in every clock period. This random variation of the amount of charge increases the noise contribution to the CT DSM output. The CT DSM performance degradation due to the pulse-width jitter highly depends on the DAC current waveform and the number of quantization bits.

In Fig. 3.20, the effect of clock jitter on a single-bit RZ, NRZ and SC DAC is shown respectively. In a single-bit RZ DAC, if the variance of the clock jitter is σ_s^2, the variance of the feedback charge error can be expressed as:

$$\sigma_{RZ}^2 = 2\sigma_s^2 I_{RZ}^2 \tag{3.26}$$

where I_{RZ} is the RZ DAC output current, and the factor 2 originates from the fact that both the rising and the falling edges are modulated by clock jitter. Assuming that the DAC duty cycle is 50%, the amount of charge transferred by the RZ DAC in one clock period is:

$$Q_{RZ}^2 = \frac{I_{RZ}^2 T_s^2}{4} \tag{3.27}$$

In a CT DSM, the clock-jitter-induced noise is oversampled, and so the SNR due to clock jitter is:

$$SNR_{RZ} = \frac{Q_{RZ}^2 OSR}{\sigma_{RZ}^2} = \frac{OSR}{8 f_s^2 \sigma_s^2} \tag{3.28}$$

For a single-bit NRZ, there are three differences with the RZ DAC case. First, because the duty cycle of the NRZ DAC is 100%, the NRZ DAC current I_{NRZ} is half the RZ DAC one I_{RZ}. For the charge transferred in one period, only one edge is influenced by clock jitter. Finally, clock jitter only introduces a charge error when there is a transition in the DAC output current. These three factors improve the NRZ DAC clock jitter performance by about 4.6dB, compared to a RZ DAC.

For a single-bit CT DSM with RZ or NRZ DAC, the clock jitter requirement is stringent. For example, in a single-bit CT DSM with 20MHz-BW, 20× OSR and 12-bit accuracy, only 0.4 and 0.67ps clock jitters can be tolerated respectively if a RZ and a NRZ DAC are used.

To relax the clock jitter requirement in single-bit CT DSMs, the switch-capacitor (SC) DAC is proposed. In a SC DAC, half a clock period is assigned for capacitor charging, and half a period is used for discharging to the loop filter, with exponentially decaying output current:

$$I_{SC}(t) = \begin{cases} I_o e^{\frac{-t}{\tau}} & nT < t < (n+0.5)T \\ 0 & (n+0.5)T < t < (n+1)T \end{cases} \tag{3.29}$$

Here I_o is the initial value of the discharge current and τ is the discharge time constant:

$$I_o = \frac{V_{ref}}{R} \tag{3.30}$$

$$\tau = RC \tag{3.31}$$

The total charge transferred by the SC DAC in one clock period is:

$$Q_{SC} = V_{ref} C \left(1 - e^{\frac{-0.5T}{RC}}\right) \tag{3.32}$$

At the end of the discharge phase, the DAC current value and the charge error variance are respectively:

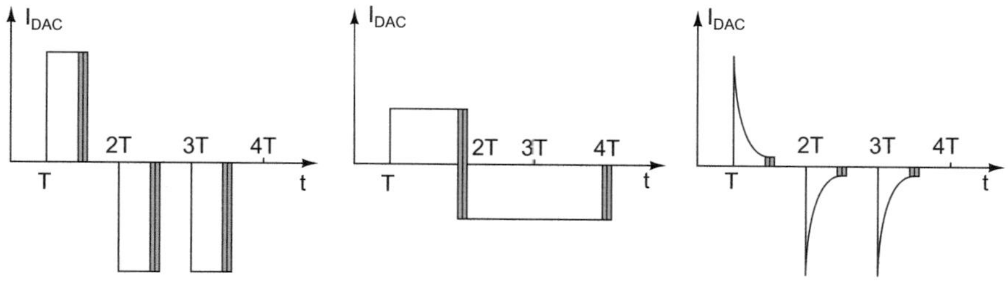

Fig. 3.20 Clock-jitter-induced noise in single-bit DACs: RZ (*left*), NRZ (*middle*) and SC (*right*)

$$I_{tail} = I_o e^{\frac{-0.5T}{\tau}} = \frac{V_{ref}}{R} e^{\frac{-0.5T}{RC}} \tag{3.33}$$

$$\sigma_{SC}^2 = \sigma_s^2 \left(\frac{V_{ref}}{R} e^{\frac{-0.5T}{RC}} \right)^2 \tag{3.34}$$

So the DSM SNR caused by clock jitter is:

$$SNR_{SC} = \frac{Q_{SC}^2 OSR}{\sigma_{SC}^2} = \frac{(RC)^2}{\sigma_s^2} \frac{\left(1 - e^{\frac{-0.5T}{RC}}\right)^2}{\left(e^{\frac{-0.5T}{RC}}\right)^2} OSR \tag{3.35}$$

If the RC time constant of the SC DAC is $0.2T$, then the clock jitter tolerance of a DSM with the above specifications is about 2.5ps.

In CT DSMs with multi-bit quantization, the clock jitter performances of the RZ and SC DACs are similar to the single-bit cases, as shown in the left and right plots in Fig. 3.21. Since their currents return to zero every period, the ratio between the delivered charge in one clock period and the charge error variance is the same for the single-bit and multi-bit cases. The only difference is that in the multi-bit case the modulator overload level is higher and thus other nonidealities are relatively lower compared to the peak signal and the corresponding clock-jitter-induced error.

Fortunately, if a multi-bit NRZ DAC is applied, the clock jitter performance can be improved by $(2^B - 1)^2\times$ compared to the single-bit case, where B is the number of quantization bits. The reason is that in a NRZ DAC, the current doesn't return to zero every period and the clock-jitter-induced noise is proportional to the current transition step (middle in Fig. 3.21), which is inversely proportional to the number of quantization steps. As a result, the combination of multi-bit quantization together with a NRZ DAC is an effective solution to clock jitter, at the expense of inter-symbol interference.

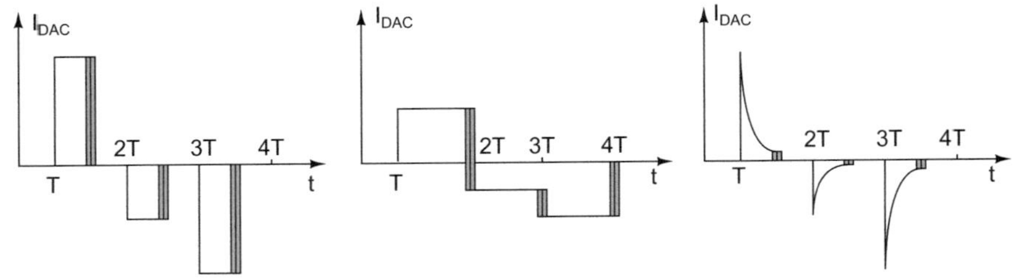

Fig. 3.21 Clock-jitter-induced noise in multi-bit DACs: RZ (*left*), NRZ (*middle*) and SC (*right*)

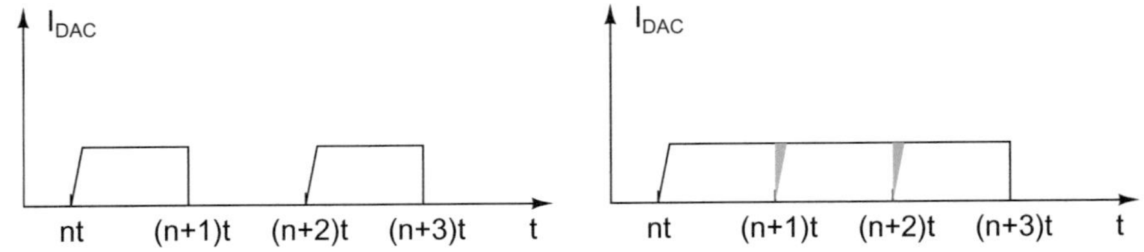

Fig. 3.22 Inter-symbol inteference (ISI) of a NRZ DAC (*right*); the error is marked in *darkgray*

3.4.2.2 Inter-symbol Interference

In CT DSMs, another nonideality of NRZ DACs is the inter-symbol interference (ISI), which is caused by the output current's unequal rising and falling edges. With unequal edges, the amount of charge transferred by one DAC cell not only depends on the number of codes "1", but also on the data sequence. In Fig. 3.22, assuming that the DAC current rise time is larger than the fall time, on the left the DAC input sequence is "101" and on the right it is "111". The ratio between the transferred charge amounts in these cases is not exactly 2/3; the error is marked in darkgray on the right. This sequence-dependency causes distortion tones in the CT DSM output spectrum and degrades the modulator SNDR. Furthermore, this degradation can not be shaped by DEM. One solution to the ISI is to adopt RZ and SC DACs, since their output current returns to zero every period, hence their transferred charge in one period is not related to the data sequence, eliminating the ISI issue.

3.4.2.3 Multi-bit DAC Mismatch

Single-bit DACs, since there are only two output levels, are intrinsically linear. However in a real implementation of a multi-bit DSM, because of the systematic and random variations of the component values, the transferred amounts of charge of different feedback DAC cells are not identical. One result of the multi-bit DAC mismatch is the harmonic distortion in the DSM output. Together with the distortion, more noise is introduced into the signal bandwidth during the DAC sampling. The

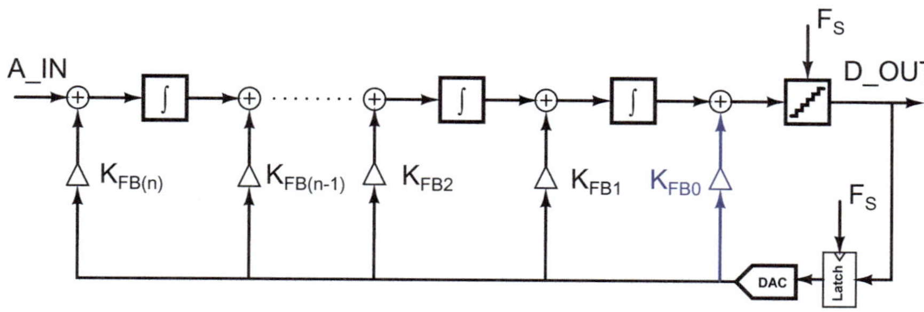

Fig. 3.23 The zero-order feedback path for ELD compensation in feedback-type DSMs

solutions for the multi-bit DAC mismatch have been briefly addressed in the last section.

3.4.3 *Quantizer Nonidealities and Modeling*

3.4.3.1 Quantizer Delay

In a CT DSM, most nonidealities of the quantizer are not so critical as those of the loop filter and feedback DAC, because they are suppressed by the preceding gain stage. However, the quantizer generates most of the excess loop delay (ELD), and the ELD will change the modulator NTF or even make the feedback loop unstable. Generally speaking, higher-order CT DSMs are more sensitive to ELD than lower-order ones. In quantizer design, the delay is determined by its time constant τ, and a smaller delay always corresponds to a larger power consumption.

To absorb the quantizer delay, usually one or half a clock period is assigned for the quantizer settling, which is executed by inserting a resampling FF between the quantizer and the feedback DAC. At the same time, to keep the NTF the same, a zero-order feedback path is added, as shown in Fig. 3.23. If the modulator is of the feedback type and the quantizer is not current mode, one extra adder is needed. To remove the extra analog adder, the second hybrid topology in Fig. 3.13 can be adopted, and a digital differentiator together with an extra DAC are used to feed back the digitally differentiated derivative signal to the last integrator, as shown in Fig. 3.24. By doing this, the digital differentiator and the analog integrator form the zero-order feedback path for the ELD compensation.

3.4.3.2 Multi-bit Quantizer Input Parasitic Capacitor

In high-bandwidth high-accuracy CT DSMs, multi-bit quantization is indispensable and the sampling frequency is very high. The exponentially-increasing number of

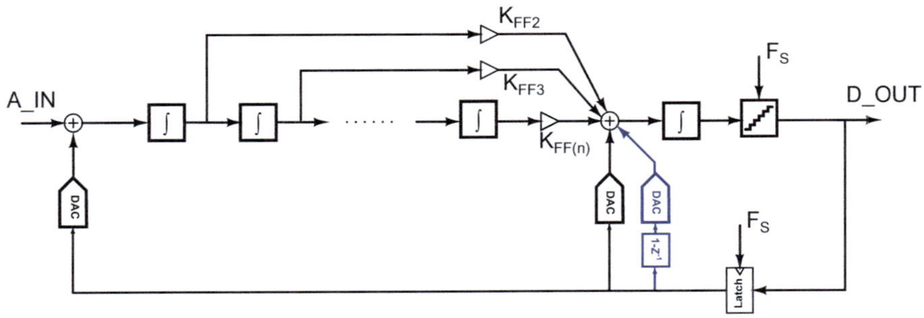

Fig. 3.24 The ELD compensation scheme without extra analog adder: the zero-order feedback path is composed by a digital differentiator and an analog integrator

comparators results in a considerable quantizer input parasitic capacitance. This parasitic capacitance together with the output impedance of the preceding stage generates an extra pole. With such a pole, the order of the CT DSM is increased by 1 and extra phase shift (time delay) is introduced.

To avoid performance degradation or instability due this extra pole, the output impedance of the preceding stage should be designed to be low enough to push the extra pole much higher than the sampling frequency. Another solution is to use the quantizer parasitic capacitance as part of a passive capacitive adder.

3.5 Conclusions

This chapter has provided the basics of CT DSMs, including the operation principle, the structural options, the modelling and the potential solution for building block nonidealities. By applying oversampling and noise shaping techniques, the DSM ADC is a high-accuracy ADC architecture without high-accuracy components. For the DSM ADC architecture, there are different options with different design aspects: discrete-time or continuous-time; single-bit or multi-bit quantization; 1st-order or higher-order noise shaping; single-loop or MASH structure; feedforward, feedback or hybrid structure. The combinations of these choices result in numerous candidate implementations. They have to be investigated and compared to each other to find the optimized structure for the required specifications in any given application. The source and the modeling of the important CT DSM building block nonidealities has also been given in this chapter. With these the modulator performance can be predicted and the specifications for the different building blocks can be determined at behavioral level, shortening the time to market and cutting the design cost.

Based on these design considerations, two CT DSM ADC designs will be presented in Chap. 5. In the next chapter, the recently developed time-domain quantization and VCO-based A/D conversion will be discussed, as a substitute for traditional DSM ADCs.

References

1. R. Schreier, G.C. Temes, *Understanding Delta-Sigma Data Converters* (Wiley, New York, 2005)
2. S.R. Norsworthy, R. Schreier, G.C. Temes, *Delta-Sigma Data Converters: Theory, Design, and Simulation* (IEEE Press, IEEE Circuit and Systems Society, 1997)
3. J.A. Cherry, W.M. Snelgrove, *Continuous-time Delta-Sigma Modulators for High Speed A/D Conversion: Theory* (Kluwer Academic Publishers, Practice and Fundamental Performance Limits, 1999)
4. M. Dessouky, A. Kaiser, Very low-voltage digital-audio modulator with 88-dB dynamic range using local switch bootstrapping. IEEE J. Solid-State Circuits **36**(3), 349–355 (2001). Mar
5. J. Crols, M. Steyaert, Switched-opamp: an approach to realize full CMOS switched-capacitor circuits at very low power supply voltages. IEEE J. Solid-State Circuits **29**(8), 936–942 (1994). Aug
6. S. Yan, E. Sanchez-Sinencio, A continuous-time sigma-delta modulator with 88-dB dynamic range and 1.1-MHz signal bandwidth. IEEE J. Solid-State Circuits **39**(1), 75–86 (2004)
7. M. Ortmanns, F. Gerfers, Y. Manoli, A continuous-time modulator with reduced sensitivity to clock jitter through SCR feedback. IEEE Trans. Circuits Syst. I: Regul. Pap. **52**(5), 875–884 (2005). May
8. P. Shettigar, S. Pavan, Design techniques for wideband single-bit continuous-time modulators with FIR feedback DACs. IEEE J. Solid-State Circuits **47**(12), 2865–2879 (2012). Dec
9. G. Mitteregger, C. Ebner, S. Mechnig, T. Blon, C. Holuigue, E. Romani, A 20-mW 640-MHz CMOS continuous-time ADC with 20-MHz signal bandwidth, 80-dB dynamic range and 12-bit ENOB. IEEE J. Solid-State Circuits **41**(12), 2641–2649 (2006). Dec
10. F.M. Gardner, A transformation for digital simulation of analog filters. IEEE Trans. Commun. **34**(7), 676–680 (1986). July
11. S.R. Norsworthy, Effective dithering of Sigma-Delta modulators, in *Proceedings International Symposium on Circuits and Systems*, 1992, pp. 1304–1307
12. W.L. Lee, A novel higher-order interpolative modulator topology for high resolution oversampling A/D converters, M.S. thesis, MIT, Cambridge, MA, 1987
13. M. Rebeschini, N.R. van Bavel, P. Rakers, R. Greene, J. Caldwell, J.R. Haug, A 16-b 160-kHz CMOS A/D converter using sigma-delta modulation. IEEE J. Solid-State Circuits **25**(2), 431–440 (1990). Apr
14. L.J. Breems, R. Rutten, G. Wetzker, A cascaded continuous-time $\Delta\Sigma$ modulator with 67-dB dynamic range in 10-MHz bandwidth. IEEE J. Solid-State Circuits **39**(12), 2152–2160 (2004). Dec
15. T. Leslie, B. Singh, An improved sigma-delta modulator architecture, in *Proceedings of the IEEE International Symposium on Circuits and Systems (ISCAS)*, pp. 372–375, May 1990
16. A. Gharbiya, D.A. Johns, A 12-bit 3.125 MHz bandwidth 0-3 MASH Delta-Sigma modulator. IEEE J. Solid-State Circuits **44**(7), 2010–2018 (2009)
17. V. Srinivasan, V. Wang, P. Satarzadeh, B. Haroun, M. Corsi, A 20mW 61dB SNDR (60MHz BW) 1b 3rd-order continuous-time Delta-Sigma modulator clocked at 6GHz in 45nm CMOS, in *2012 IEEE International on Solid-State Circuits Conference Digest of Technical Papers (ISSCC)*, pp. 158–160, Feb 2012

18. P. Shettigar, S. Pavan, A 15mW 3.6GS/s CT-ADC with 36MHz bandwidth and 83dB DR in 90nm CMOS, in *2012 IEEE International on Solid-State Circuits Conference Digest of Technical Papers (ISSCC)*, pp. 156–158, Feb 2012
19. Y. Geerts, M.S.J. Steyaert, W. Sansen, A high-performance multibit CMOS ADC. IEEE J. Solid-State Circuits **35**(12), 1829–1840 (2000). Dec
20. J.G. Kauffman, P. Witte, J. Becker, M. Ortmanns, An 8.5 mW continuous-time modulator with 25 MHz bandwidth using digital background DAC linearization to achieve 63.5 dB SNDR and 81 dB SFDR. IEEE J. Solid-State Circuits **46**(12), 2869–2881 (2011)
21. J. Silva, U. Moon, J. Steensgaard, G.C. Temes, Wideband low-distortion delta-sigma ADC topology. Electron. Lett. **37**(12), 737–738 (2001). June
22. L. Yao, M. Steyaert, W. Sansen, A 1-V, 1-MS/s, 88-dB sigma-Delta modulator in 0.13-μm digital CMOS technology, in *2005 Symposium on VLSI Circuits, 2005*. Digest of Technical Papers, June 2005, pp. 180–183
23. F. Gerfers, M. Ortmanns, *Continuous-Time Sigma-Delta A/D Conversion: Fundamentals, Performance Limits and Robust Implementations* (Springer, Berlin Heidelberg, 2006)

Chapter 4
VCO-Based ADCs

4.1 Introduction

With CMOS technology scaling, the analog and mixed-signal circuits face more and more design challenges and suffer a lot in accuracy. At the same time, digital circuits benefit from technology scaling in terms of improved timing accuracy and reduced power consumption. As a result, more and more quantization and A/D conversions are transferred from the analog and the voltage domains to the digital and the time domains, by using ring VCOs and pulse-width modulation (PWM). Four different kinds of VCO-based quantizer have been developed for power-efficient conversion, with a highly-digital structure and intrinsic 1st-order noise shaping. In a VCO-based counting quantizer, the VCO maximum frequency is higher than half of the sampling frequency and a counter is required. By reducing the VCO maximum frequency, the counter can be replaced by a FF. In the frequency-type VCO-based quantizer, there is an implicit DEM on its digital output. However, the VCO voltage-to-frequency (V-F) nonlinearity is the performance bottleneck. For the phase-type quantizer, the VCO nonlinearity is not a limitation, but there is no implicit DEM and the nonlinearity of the multi-bit feedback DAC restricts the ADC SNDR. In closed-loop DSMs with a frequency-type VCO-based quantizer, the modulator linearity can be improved by analog loop filter suppression and input swing reduction. If the VCO-based quantizer is applied in an open-loop ADC, there is only 1st-order noise shaping, and the quantizer nonlinearity can be solved by digital LUT-based calibration (both foreground and background) or by PWM precoding.

This chapter provides an overview and a discussion about the design of recently developed VCO-based ADCs. First, the structures and the circuits of four VCO-based quantizers are presented; their merits and shortcomings are also explored and compared. Then three closed-loop DSM ADC structures with VCO-based quantizers are shown, focusing on how to solve the frequency-type quantizer nonlinearity. Finally, three open-loop VCO-based ADC architectures are also introduced, together with their linearity enhancement techniques: digital LUT-based calibration and PWM precoding.

X. Xing et al., *Design of Power-Efficient Highly Digital Analog-to-Digital Converters for Next-Generation Wireless Communication Systems*, Signals and Communication Technology, https://doi.org/10.1007/978-3-319-66565-8_4

4.2 VCO-Based Quantizers

In SoC design, the analog and mixed-signal circuits are expected to follow the CMOS technology scaling, which is mainly optimized for digital circuits. With process scaling, the transistor gate length and parasitic capacitance become smaller, while the oxide capacitance density increases. As a result, the operating frequency of CMOS transistor increases. However, analog and mixed-signal circuits face several design challenges, which have been introduced in Chap. 1 and need to be addressed to accommodate the developments in SoC design. On the other hand, the gate delay and the power consumption of digital circuits are continuously decreasing, as the timing accuracy in the time domain is improved. Based on these facts, more and more ADC designs are executed in the time and the digital domains.

Both PWM and VCO are common blocks used for the transformation from the voltage domain to the time domain. They have the same characteristic as a comparator: the input amplitude is continuous while the output amplitude is discrete. which implies that with proper arrangement, they can be used as time-domain quantizer to replace the voltage-domain quantizer (comparator and flash ADC). The major difference is that their information is stored in the edge positions, but not in the voltage levels. Until now several VCO-based quantizer structures have been proposed, some of them using a counter and register to process the VCO output phase, including a single-phase and multi-phase quantizer. In the frequency- and phase-type VCO-based quantizers, a register and XOR gate are used for the information extraction.

4.2.1 Single-Phase Counting VCO-Based Quantizer

The first VCO-based quantizer [1] has been proposed in 1977 for power applications. It comprises a single-phase VCO, a counter and a sampling register, as shown on the left in Fig. 4.1. The oscillator frequency is controlled continuously by the analog input voltage. With a rising edge of the reference clock, the counter is reset and begins to accumulate the number of VCO output state transitions. After one clock period, the counting result is sampled by the register and one full process is finished. The quantizer digital output D_{OUT} is proportional to the VCO frequency, and thus to the analog input voltage A_{IN}. It is clear that the VCO maximum frequency is exponentially proportional to the required number of bits, making this approach not practical.

4.2.2 Multi-phase Counting VCO-Based Quantizer

To improve the quantization accuracy and to reduce the VCO maximum frequency, a multi-phase counting VCO-based quantizer [2] has been developed, as illustrated on the right in Fig. 4.1. A multi-phase VCO, which is usually implemented by a ring

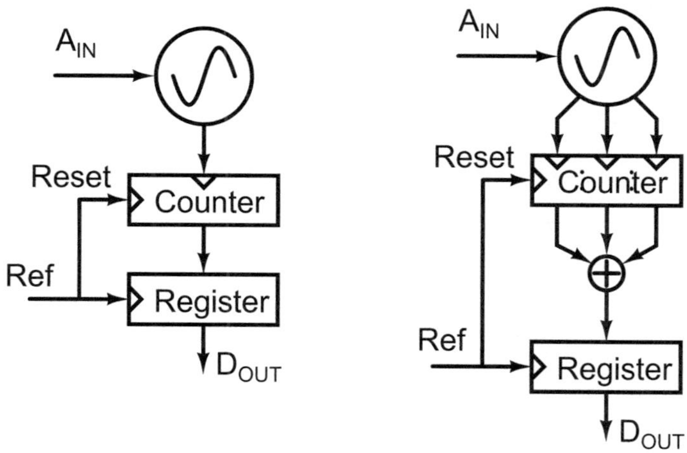

Fig. 4.1 Single-phase (*left*) and multi-phase (*right*) counting quantizer

VCO, is modulated by the analog input voltage. Each output phase of the VCO is used to trigger one counter. Once the counter is reset by the clock edge, the numbers of state transitions for all VCO phase outputs are counted and all counter outputs are summed together. With the next clock edge, the summation output is sampled and all counters are reset. With more than one phase output, the VCO maximum frequency of this structure is reduced compared to that of the single-phase counting VCO-based quantizer. However, the design of the counters and the adder are more complex. In both the multi-phase and the single-phase counting quantizers, there is an inherent 1st-order noise shaping for the quantization noise [3].

4.2.3 Frequency-Type VCO-Based Quantizer

If the number of VCO output transitions is no more than 1 in one clock period, the counter for the multi-phase counting quantizer can be replaced by a flip flop (FF), simplifying the design of the digital part. The structure and model of a widely used VCO-based quantizer (B-bit) [4, 5] is shown in Fig. 4.2. First a $\left(2^B - 1\right)$-stage ring VCO converts the input voltage into phase information, which is illustrated by the rising/falling edges of the $\left(2^B - 1\right)$ VCO phase outputs. Then each VCO phase is fed to a slice of frequency detector. A sense-amplifier flip flop (SA FF) is used to synchronize each VCO output, introducing a quantization error in the phase domain. Finally a digital differentiator (comprised of a TSPC FF and an XOR gate) processes the sampled phase and outputs the digital frequency information. The digital output actually is the number of VCO delay cells, that change their state once (from "1" to "0" or from "0" to "1") in one clock period. In this quantizer, since there is no register assigned for the transition counting, the number of state transitions of each phase

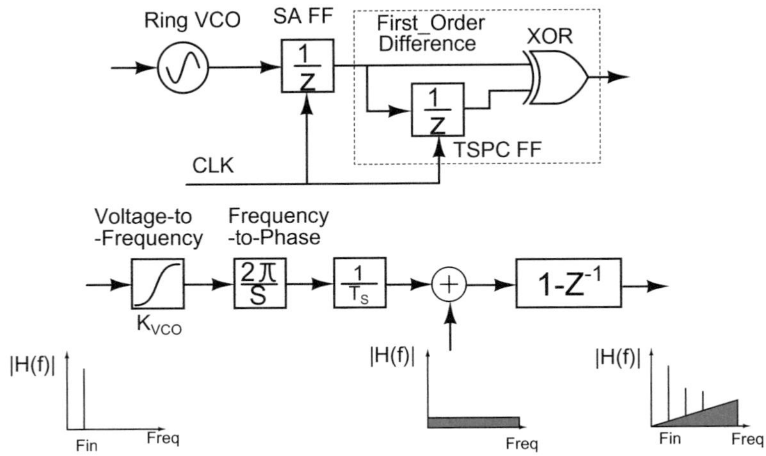

Fig. 4.2 Structure, behavioral model and illustrative spectrum of the frequency-type VCO-based quantizer

output may not be larger than 1, or else the digital frequency output is reset and the desired information is ruined. With this limitation, the VCO maximum frequency can not exceed half of the quantizer sampling frequency. A proper design is to set the VCO maximum frequency to half the sampling frequency, and the free-running frequency to one quarter of the sampling frequency. Since the VCO minimum frequency is much smaller compared to the maximum frequency, the VCO gain is:

$$K_{VCO} = \frac{f_{max} - f_{min}}{A_{in}} \approx \frac{f_s}{2A_{in}} \tag{4.1}$$

Compared to a flash ADC, the VCO-based quantizer has two extra advantages, which make it very suitable for DSM design. First, with the VCO integration and digital differentiation, the transfer function from the signal input to the output is unity, while the transfer function from the quantization error source to the output is a differentiator. Because of the VCO integration $\frac{2\pi K_{VCO}}{s}$, the quantization noise is inherently 1st-order noise-shaped. With this 1st-order noise shaping and sufficient OSR, even a single frequency-type VCO-based quantizer can reach moderate noise performance. Several examples will be presented later.

In a ring VCO, the state transition of a delay cell corresponds to a "1" at its frequency detector output. Since the delay cells change their state one by one, the frequency detectors output "1" consecutively. In other words, a DEM is intrinsically implemented with this frequency-type VCO-based quantizer, without any extra timing penalty and power consumption. If this quantizer is applied in a closed-loop DSM, then the implicit DEM relaxes the front-end DAC matching requirement greatly.

On the other hand, the relationship between the input voltage and the VCO frequency is not linear. In behavioral simulations, a polynomial can be used to model the V-F nonlinearity:

$$f = f_o + k_{vco}A_{in} + k_2A_{in}^2 + k_3A_{in}^3 + k_4A_{in}^4 \tag{4.2}$$

Such nonlinearity is not suppressed by any filtering and generates harmonic distortion in the quantizer output. Although the even-order distortion can be cancelled by a pseudo-differential topology, the 3rd-order distortion still exists and limits the quantizer linearity to about 5-bit. How to solve this nonlinearity is the key challenge to apply this quantizer in DSM designs.

In the VCO-based quantizer design, the VCO phase noise must also be considered, especially if it is used in open-loop ADCs, where there is no preceding analog loop filter suppression. Let's assume that the VCO phase noise is L at the offset frequency f_{off}, and the input signal amplitude and frequency are A_{in} and f_{in} respectively, then the peak signal-to-phase-noise-ratio of the quantizer can be expressed as [6]:

$$SPNR \approx 10log\left(\frac{(K_{VCO}A_{in})^2}{16Lf_{off}^2f_{in}}\right) \tag{4.3}$$

If the clock of the VCO-based quantizer is not clean enough, it introduces an error in the quantization through two mechanisms: the sampling uncertainty triggered by pulse delay jitter and the integration error caused by pulse width jitter. The sampling error in the VCO-based quantizer is the same as in conventional ADCs:

$$SJNR_d = -20log(2\pi f_{in}\sigma_d) \tag{4.4}$$

Here σ_d^2 is the pulse delay jitter variance and f_{in} is the input signal frequency. If the variance of the pulse width jitter is σ_w^2, then the quantizer performance is:

$$SJNR_w = 10log\left(\frac{T_s^2}{\sigma_w^2}\frac{sinc^2\left(\frac{f_{in}}{f_s}\right)}{1+2\left(\frac{f_{fr}}{K_{VCO}A_{in}}\right)^2}\right) \tag{4.5}$$

where f_{fr} is the VCO free-running frequency. In addition, the VCO-based quantizer has a $(2^B - 1) \times$ smaller probability of being metastable than a flash ADC. If the delay cells are implemented by minimum gate length transistors, then the offset voltage can reach tens of mV. Fortunately, the quantization error due to the offset is also 1st-order shaped. Finally, the VCO-based quantizer is also statistically monotonic.

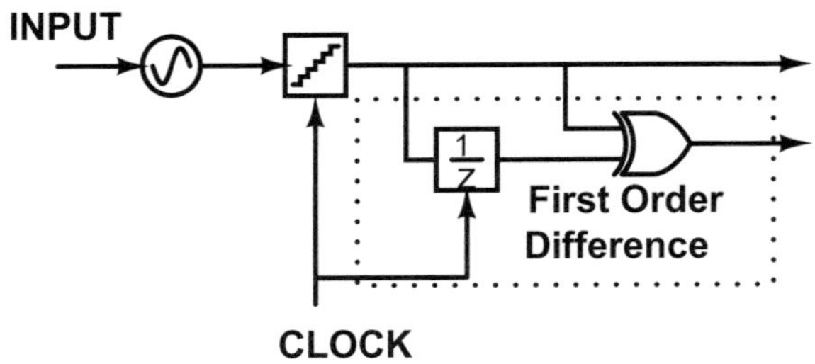

Fig. 4.3 Structure of the phase-type VCO-based quantizer; the additional digital frequency output is used for DSM ELD compensation

4.2.4 *Phase-Type VCO-Based Quantizer*

As mentioned in the last subsection, the VCO nonlinearity is the performance bottleneck in the frequency-type VCO-based quantizer. To avoid this disadvantage, a phase-type VCO-based quantizer [7, 8] has been proposed, as shown in Fig. 4.3. In this quantizer, the VCO phase and not the frequency is adopted as the digital output. To remove the common-mode information in the digital phase output, an xor operation between the VCO outputs and quadrature pulses with the VCO free-running frequency is taken after the SA FF sampling. Now the transfer function from the signal input to the output is an integration, while the function from the quantization error source to the output is unity. As a result, the intrinsic 1st-order quantization noise shaping is the same as that of the frequency-type VCO-based quantizer.

Because the signal transfer function is an integration and not unity, the phase-type VCO-based quantizer cannot be used separately as an open-loop ADC; it can only be applied in a closed-loop system. In a closed-loop DSM with a phase-type VCO-based quantizer, because of the large gain provided by the VCO integration, the VCO input voltage swing is reduced greatly, and so is its nonlinearity. Behavioral simulations show that the harmonic distortion of this quantizer can be as low as −90 dBc [8], which is enough for most DSM designs.

Another drawback of the phase-type VCO-based quantizer is that there is no implicit DEM on the quantizer digital outputs. Consequently, the DAC matching is a potential design challenge and needs to be solved by others methods.

In CT DSM design, usually a zero-order feedback path is added to compensate the ELD. To implement a zero-order feedback path, frequency detectors are also applied in the phase-type quantizer to provide the digitized frequency. For this secondary output, the intrinsic DEM still remains. Because of the reduced VCO input voltage swing, the digital frequency output is also a low-swing signal.

4.3 Closed-Loop VCO-Based DSMs

In this section, three different published closed-loop DSMs with VCO-based quantizers are presented, and the design solutions for the VCO nonlinearity are discussed.

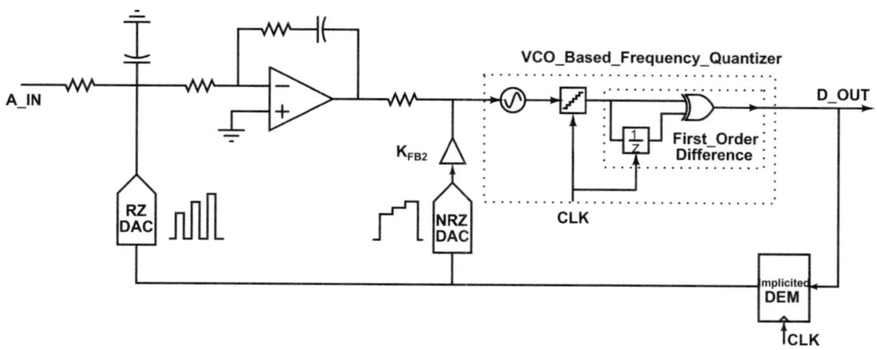

Fig. 4.4 Architecture of a 3rd-order CT DSM with frequency-type VCO-based quantizer [4]

4.3.1 DSM with Frequency-Type VCO-Based Quantizer

A closed-loop DSM [4, 5] with frequency-type VCO-based quantizer is shown in Fig. 4.4. It is a 3rd-order feedback structure with 5-bit time-domain quantization. 1st-order noise shaping comes from the VCO-based quantizer, and the other 2nd-order shaping is done by the analog loop filter. The nonlinearity of the VCO is suppressed by the preceding loop filter. In the 2nd-order loop filter, the front-end stage is implemented by a passive RC filter and the second stage by an active-RC integrator. Since the passive low-pass filter doesn't provide any gain, the quantization noise and VCO nonlinearity are suppressed by 2nd and 1st-order filters respectively. Furthermore, the resistor of the RC filter contributes to the DSM noise power and degrades its SNR. However, the passive front-end filters out the high-frequency signal component from the RZ DAC, and makes the input signal of the active-RC integrator smooth. This is beneficial for the linearity of the active integrator.

A RZ DAC is chosen as the outermost DAC based on two design considerations. First, a RZ DAC is robust to ISI. Secondly, by using a RZ DAC, more ELD can be tolerated by the modulator. As a result the CT DSM is still marginally stable, even if the inner compensation DAC is disabled. On the other hand, compared to a NRZ counterpart, the RZ DAC has double the output current and requires a larger OTA slew rate. As mentioned before, this disadvantage can be mitigated by the front-end passive filtering. Besides, the RZ DAC is much more sensitive to clock jitter, which means that a very clean clock generator is required for measurements and in application. With the implicit DEM of the frequency-type VCO-based quantizer, the RZ DAC matching is not the performance bottleneck.

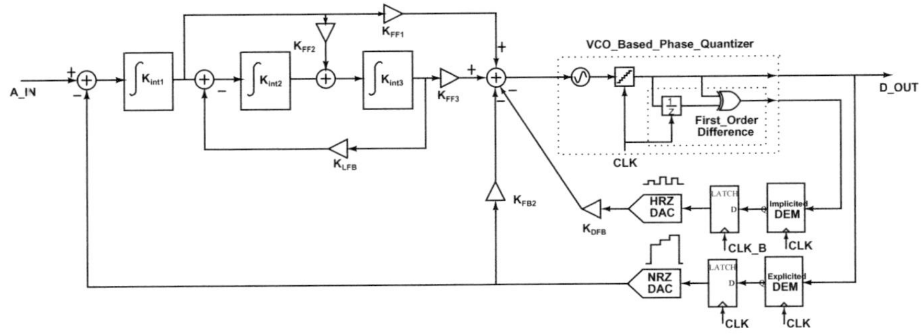

Fig. 4.5 Topology of a 4th-order CT DSM with phase-type VCO-based quantizer [8]

4.3.2 DSM with Phase-Type VCO-Based Quantizer

In the above design, even with filter suppression, the VCO nonlinearity is still considerable and the modulator SFDR is only about 75 dB. To achieve better linearity performance, the phase-type VCO-based quantizer is applied to a 4th-order 4-bit CT DSM with hybrid structure [7, 8] as illustrated in Fig. 4.5.

1st-order noise shaping arises from the VCO integration, and the other 3rd-order noise shaping is from the active-RC loop filter. Because of the VCO integration, the VCO input swing is so small that its V-F characteristic can be considered to be very linear. Hence the VCO nonlinearity never limits the DSM performance.

Without quantizer implicit DEM, the front-end DAC linearity is a primary design challenge, which is solved by an explicit DEM. Since only half a clock period is assigned for the DEM timing, high-speed dynamic logic is used to implement the DEM, with a large power consumption penalty. In the outermost NRZ DAC design, a degeneration resistor is use to reduced the DAC output noise by a factor of $(g_m R)^2$, where g_m is the transconductance of the current source transistor in the DAC cell, and R is the degeneration resistor. For the inner NRZ DAC, since its nonidealities are shaped by the 3rd-order analog filter, its noise and matching requirements are relaxed to save chip area. For the RZ DAC, the low input swing and implicit DEM reduce its area further. To save power consumption, a passive resistive adder is adopted to sum the integrator voltages and DAC currents.

4.3.3 DSM with Residual-Cancelling VCO-Based Quantizer

Based on the frequency-type VCO-based quantizer, the residual-cancelling concept has been proposed to solve the VCO nonlinearity. In Fig. 4.6, a 2nd-order CT DSM with a residual-cancelling VCO-based quantizer [9, 10] is presented. The residual-cancelling VCO-based quantizer (in the dashed box) comprises a flash ADC, a subtractor, a DAC and a frequency-type VCO-based quantizer. First, the quantizer analog

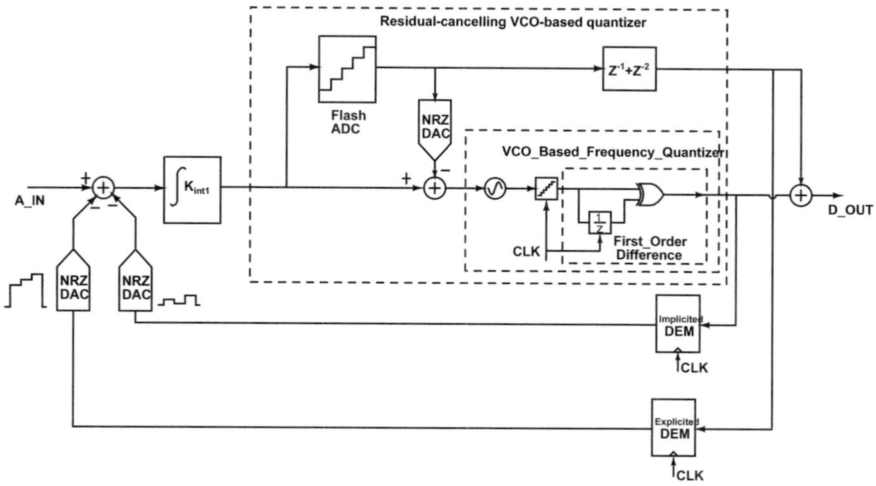

Fig. 4.6 Structure of a 2nd-order CT DSM with residual-cancelling VCO-based quantizer [10]

input is converted into digital code by the flash ADC. Then, by using a DAC and a subtractor, the residual error of the flash conversion is generated, which is processed further by the frequency-type VCO-based quantizer. The outputs of the flash ADC and the VCO-based quantizer are summed to create the whole quantizer digital output. From the flash ADC output to the overall quantizer output, there are two paths: the digital path and the path composed by the DAC, the subtractor and the VCO-based quantizer. If the gains of these two paths are equal, then the flash ADC noise and distortion are cancelled. With the low-swing residual error as input, the frequency-type VCO-based quantizer is very linear, not limiting the DSM performance. In the frequency-type VCO-based quantizer, the quantization is in the time domain and not in the voltage domain. So even with a low voltage swing input, all the quantization levels are used and the quantization error is small.

In the design of Fig. 4.6, 1st-order noise shaping is provided by the VCO-based quantizer and another 1st-order shaping comes from the active-RC integrator. The analog loop filter also suppresses all nonidealities of the residual-cancelling VCO-based quantizer. For the multi-bit DAC feedbacking the flash ADC output, its digital input has full swing and no intrinsic DEM, so an explicit DEM is required to ease its matching. The matching requirement of the DAC connected to the VCO-based quantizer is very loose because of two reasons: first this DAC's input swing is much smaller than the full swing; secondly there is implicit DEM on its digital input.

4.4 Open-Loop VCO-Based ADCs

In this section, three published open-loop 1st-order VCO-based ADCs are presented, with different solutions to address the VCO nonlinearity. Without global feedback, these ADCs have no stability issue like closed-loop DSMs.

4.4.1 VCO-Based ADC with Background Digital Calibration

In Fig. 4.7, the architecture of a 1st-order open-loop Delta-Sigma ADC [11, 12] is presented, it is a two-path structure, but only one path of the ADC is shown here. In each path, to cancel 2nd-order harmonic distortion, a pseudo-differential ADC topology is utilized. First the fully differential transconductance converts the input voltage into current, which is fed to the pseudo-differential current-controlled ring oscillator (ICO). Each ICO phase output is sampled and processed by the digital differentiator to obtain the digital frequency information. Although the V-to-I conversion can be very linear, because of the nonlinearity of the capacitor of the ICO delay cell, the I-to-F conversion is still nonlinear. This nonlinearity determines the VCO-based ADC performance directly. It is removed by the nonlinearity correction block, which is explained in the next paragraph. To remove the idle tone in this 1st-order Delta-Sigma ADC, a dithering signal is generated by the linear feedback shift register (LFSR) and added to the ICO inputs. The polarities of the dithering signal added to the two paths are opposite, and by adding the two path outputs, the dither signal is cancelled at the output of the whole ADC.

To eliminate the VCO distortion, a background digital nonlinearity correction block is designed. The correction coefficients are as shown in Fig. 4.8. To measure the ICO 2nd- and 3rd-order distortion coefficients, a replica single-ended path is added and a calibration sequence $t_1[n] + t_2[n] + t_3[n]$ is used as its input. Here $t_i[n]$ (i = 1, 2, 3) is a 2-level, independent, zero mean, pseudo sequence generated by another low-speed LFSR. In one calibration cycle, the replica path output is correlated with three 2-level sequences respectively: $t_1[n]$, $t_1[n] + t_2[n]$ and $t_1[n] + t_2[n] + t_3[n]$, and

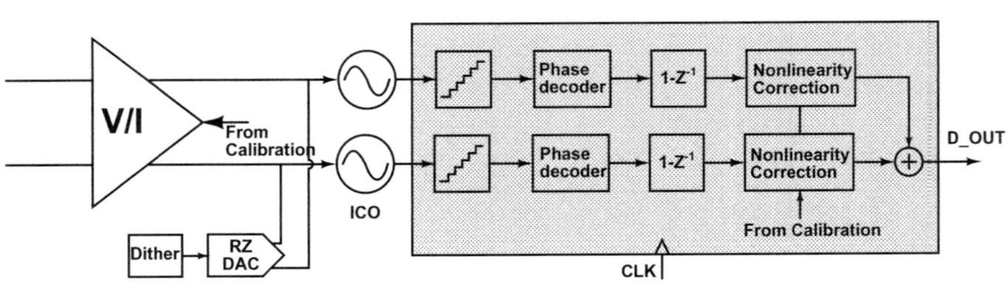

Fig. 4.7 Architecture of a 1st-order open-loop Delta-Sigma ADC with VCO nonlinearity calibration (only one path is shown here) [11]

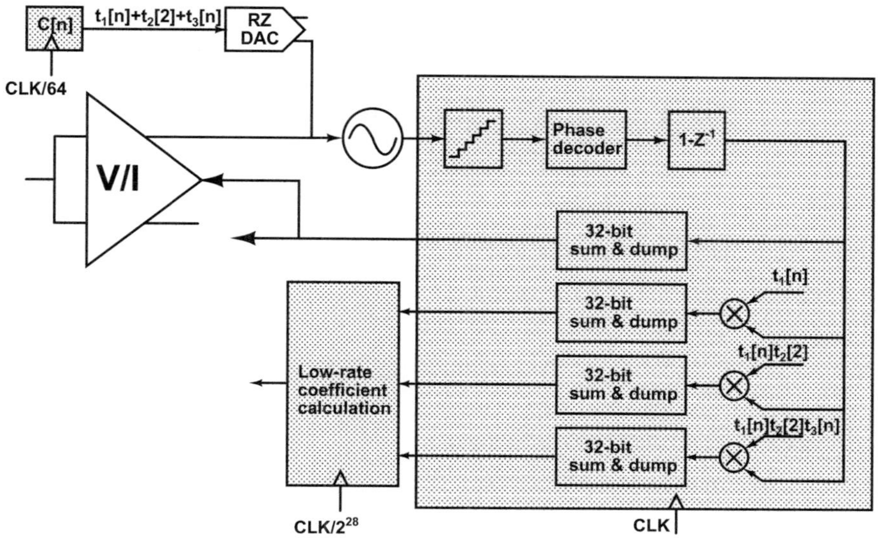

Fig. 4.8 Concept of the VCO nonlinearity background digital calibration used in Fig. 4.7 [11]

2^{28} correlation outputs are summed together to calculate the distortion coefficients. With such digital calibration, the VCO-based ADC SFDR is improved from 50 dB to better than 80 dB.

4.4.2 VCO-Based ADC with Counting and Foreground Digital Calibration

An alternative open-loop VCO-based ADC with foreground digital foreground calibration [13] is shown in Fig. 4.9. In this design, the VCO maximum frequency is much higher than the sampling clock. To avoid that the VCO output phase transition information is ruined by the next round of transitions, one delay cell output is connected to an asynchronous counter. The bit-length of the counter depends on the ratio of the VCO maximum frequency and the sampling frequency. The counter output is sampled and handled by a multi-bit digital differentiator as the coarse output of the VCO-based ADC. The fine ADC output is the result of frequency detectors processing all the VCO output phases.

With such a coarse-fine ADC architecture, the power consumption can be reduced based on the tradeoff between different design parameters. The global optimization is based on combinations for different values of the OSR and the number of quantization bits B. In the local optimization, different numbers of quantization bits comes from the asynchronous counter and from the frequency detectors. With a M-bit counter, the VCO maximum frequency can be as high as $(2^M - 1)\times$ the sampling frequency.

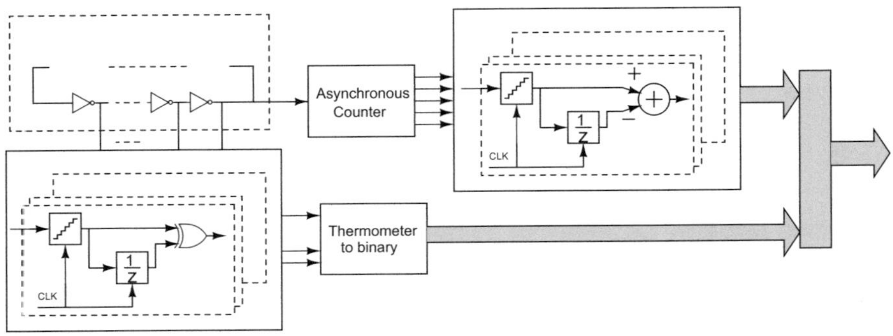

Fig. 4.9 Structure of an open-loop Delta-Sigma ADC with coarse and fine quantizations and VCO nonlinearity calibration [13]

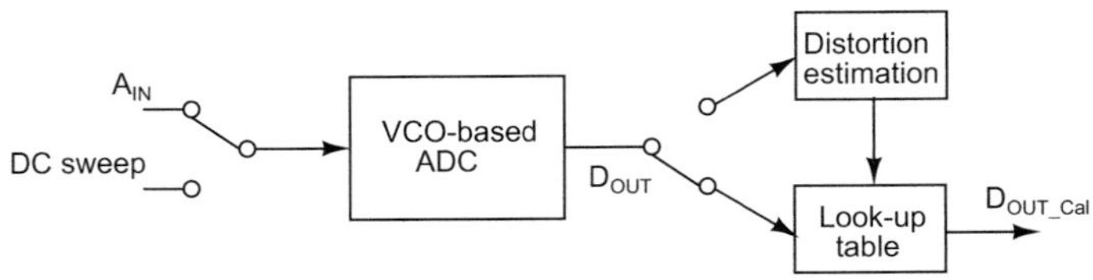

Fig. 4.10 Concept of the foreground digital calibration used in Fig. 4.9 [13]

Since $(B - M)$-bit quantization is from the frequency detectors, a $\left(2^{B-M} - 1\right)$-stage VCO and $\left(2^{B-M} - 1\right)$ frequency detectors are required. If the maximum speed of each block is known and the power estimation models are developed for the different blocks, the parameter combination with the lowest power consumption can be found.

Although a pseudo-differential architecture is applied, the 3rd-order and the residual 2nd-order harmonic distortions limit the VCO-based ADC performance to about 5 bit. To remove these distortions, a LUT-based digital foreground calibration is developed; the concept is illustrated in Fig. 4.10. In the calibration mode, 13 DC levels are applied to the ADC input and the digital outputs are processed by a distortion estimation algorithm. The algorithm fits a straight line to the measured nonlinear V-F transfer function, with a least squares method. The error between the line and the measured function is calculated, interpolated and stored in a LUT. In the normal operation mode, the VCO-based ADC output is corrected by the coefficients stored in the LUT.

4.4.3 VCO-Based ADC with PWM Precoding

In the above two open-loop VCO-based ADCs, the VCO nonlinearity is eliminated by digital calibration. Another method used for VCO-based ADC linearity improvement is PWM precoding [14, 15]. The block diagram of an open-loop VCO-based

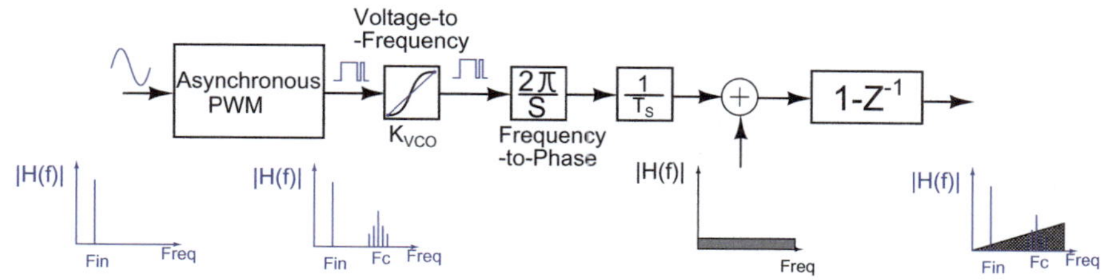

Fig. 4.11 Architecture and illustrative signals of an open-loop VCO-based ADC with PWM precoding [15]

Fig. 4.12 Implementation of an asynchronous PWM with hysteresis single-bit quantizer [16]

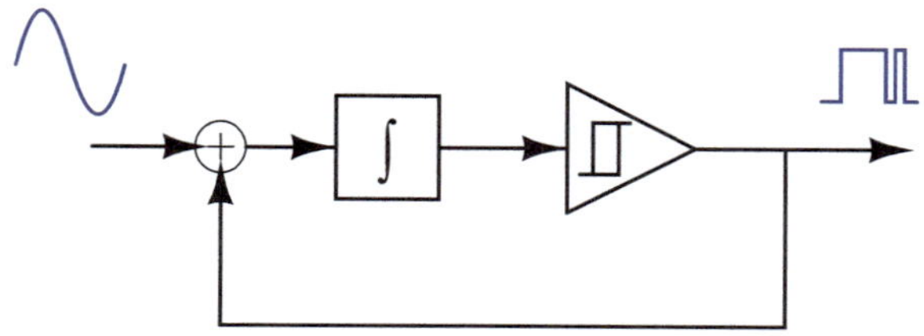

ADC with PWM precoding is presented in Fig. 4.11. The ADC is composed of an asynchronous PWM modulator and a frequency-type VCO-based quantizer. First the analog input signal is modulated into a PWM signal. Because of the 2-level characteristic of the PWM signal, the VCO only oscillates at two frequencies, and the VCO V-F transfer function is intrinsically linear.

One implementation of the asynchronous PWM modulator is shown in Fig. 4.12. It is similar to a 1st-order single-bit DSM, except that there is no clock in the modulator and hysteresis is intentionally added to the comparator. Since there is no sampling in the PWM modulator, there is no quantization noise. However, it introduces intrinsic 3rd-order distortion and some high-frequency signal components around the oscillation frequency. The 3rd-order distortion amplitude depends on the input signal frequency. The worst case is that the 3rd-order harmonic is located at the signal bandwidth, and the HD3 can then be estimated as [16]:

$$HD_3 = \frac{\pi^2}{216} \frac{M^2}{\left(1 - \frac{M^2}{2}\right)^2} \frac{1}{OCR^2} \tag{4.6}$$

Here M is the modulation depth (the ratio between the input signal amplitude and the reference value of the feedback path) and OCR is the overcycling ratio (the ratio between the self-oscillation frequency f_c and the input signal bandwidth).

4.5 Conclusions

To be compatible with SoC design in nanoscale CMOS technologies, highly-digital time-domain ADCs gain more and more attention and several novel ADC architectures have been proposed. In this chapter, recently developed time-domain quantizers and A/D converters have been presented and discussed. Based on the number of VCO output phase transitions, two kinds of VCO-based counting quantizer have been developed. If the VCO maximum frequency is less than half the sampling frequency, the counter can be replaced by a simple FF and two types of VCO-based quantizer are proposed. With such highly-digital VCO-based quantizers, not only sampling and quantization are done, but also 1st-order noise shaping and intrinsic DEM (for the frequency-type VCO-based quantizer) are provided. The signal transfer function of the phase-type VCO-based quantizer is an integrator, which can only used in closed-loop DSM designs. For frequency-type VCO-based quantizers, the VCO nonlinearity is the major design challenge when designing moderate- or high-accuracy ADCs. One solution is to apply this quantizer in closed-loop DSMs and to suppress the VCO nonlinearity by the analog loop filter. Another solution is to use the VCO-based quantizer for residual signal conversion, resulting into the residual-cancelling ADC structure.

To further digitalize the ADC design, three open-loop ADCs with VCO-based quantization have been introduced. Without global feedback, these ADCs are very stable and their quantization noise is intrinsically 1st-order shaped. In these ADC designs, if there is a counter assigned for one VCO phase output, then less VCO stages are required and the VCO frequency can be set higher. One popular solution to the VCO nonlinearity is the digital LUT-based calibration, using either a random sequence or a DC voltage as calibration input. Another architecture is using a PWM-modulated signal and not an analog input signal to drive the frequency-type VCO-based quantizer; the PWM two-points characteristic eliminates the VCO nonlinearity issue. However, the PWM will introduce inherent 3rd-order harmonic distortion.

In Chaps. 6, 7 and 8, we will introduce three different VCO-based ADCs (both closed-loop ones and open-loop one) with excellent power efficiencies that mitigate the VCO nonlinearity.

References

1. V.B. Boros, A digital proportional integral, and derivative feedback controller for power conditioning equipment, in *IEEE Power Electronics Specialists Conference* (1977), pp. 135–141
2. J.P. Hurrell, D.C. Pridmore-Brown, A.H. Silver, Analog-to-digital conversion with unlatched SQUID's. IEEE Trans. Electron Devices **27**(10), 1887–1896 (1980)
3. M. Hovin, A. Olsen, T.S. Lande, C. Toumazou, Delta-sigma modulators using frequency-modulated intermediate values. IEEE J. Solid-State Circuits **32**(1), 13–22 (1997)
4. M.Z. Straayer, M.H. Perrott, A 12-bits, 10-MHz bandwidth, continuous-time Delta-Sigma ADC with a 5-bits, 950-MS/s VCO-based quantizer. IEEE J. Solid-State Circuits **43**(4), 805–814 (2008)

5. M.Z. Straayer, M.H. Perrott, A 10-bit 20-MHz 38-mW 950-MHz CT $\Delta\Sigma$ ADC with a 5-bit noise-shaping VCO-based quantizer and DEM circuit in 0.13-μm CMOS, in *VLSI Symposium Dig.* (2007), pp. 246–247
6. J. Kim, T.K. Jang, Y.G. Yoon, S.H. Cho, Analysis and design of voltage-controlled oscillator based analog-to-digital converter. IEEE Trans. Circuits Syst. I Regul. Pap. **57**(1), 18–30 (2010)
7. M. Park, M.H. Perrott, A 0.13 μm CMOS 78 dB SNDR 87 mW 20 MHz BW CT $\Delta\Sigma$ ADC with VCO-based integrator and quantizer, in *2009 IEEE International Solid-State Circuits Conference - Digest of Technical Papers, ISSCC* (2009), pp. 170–171
8. M. Park, M.H. Perrott, A 78 dB SNDR 87 mW 20 MHz bandwidth continuous-time $\Delta\Sigma$ ADC with VCO-based integrator and quantizer implemented in 0.13 μm CMOS. IEEE J. Solid-State Circuits **44**(12), 3344–3358 (2009)
9. K. Reddy, Rao S., R. Inti, B. Young, A. Elshazly, M. Talegaonkar, P.K. Hanumolu, A 16 mW 78 dB-SNDR 10 MHz-BW CT-DSM ADC using residue-cancelling VCO-based quantizer, in *2012 IEEE International Solid-State Circuits Conference Digest of Technical Papers (ISSCC)* (2012), pp. 152–154
10. K. Reddy, Rao S., R. Inti, B. Young, A. Elshazly, M. Talegaonkar, P.K. Hanumolu, A 16-mW 78-dB SNDR 10-MHz BW CT $\Delta\Sigma$ ADC using residue-cancelling VCO-based quantizer. IEEE J. Solid-State Circuits, **47**(12), 2916–2927 (2012)
11. G. Taylor, I. Galton, A mostly-digital variable-rate continuous-time ADC $\Delta\Sigma$ modulator, in *2010 IEEE International Solid-State Circuits Conference Digest of Technical Papers (ISSCC)* (2010), pp. 298–299
12. G. Taylor, I. Galton, A mostly-digital variable-rate continuous-time Delta-Sigma modulator ADC. IEEE J. Solid-State Circuits **45**(12), 2634–2646 (2010)
13. J. Daniels, W. Dehaene, M. Steyaert, A. Wiesbauer, A 0.02 mm^2 65 nm CMOS 30 MHz BW all-digital differential VCO-based ADC with 64 dB SNDR, in *2010 IEEE Symposium on VLSI Circuits (VLSIC)* (2010), pp. 155–156
14. L. Hernandez, S. Paton, E. Prefasi, VCO-based sigma-delta modulator with PWM precoding. Electron. Lett. **47**(10), 588–589 (2011)
15. S. Rao, B. Young, A. Elshazly, W. Yin, N. Sasidhar, P.K.Hanumolu, A 71 dB SFDR open-loop VCO-based ADC using 2-level PWM modulation, in *2011 Symposium on VLSI Circuits (VLSIC)* (2011), pp. 270–271
16. P. Gao, X. Xing, J. Craninckx, G. Gielen, Design of an intrinsically-linear double-VCO-based ADC with 2nd-order noise shaping, in *2012 Design, Automation and Test in Europe Conference and Exhibition (DATE)* (2012), pp. 1215–1220

Chapter 5
CT DSM ADCs with VCO-Based Quantization

5.1 Introduction

Ever higher data rates in both wireless and wireline communications lead to an ongoing increase in signal bandwidths. In receiver chains, the variable gain amplifiers (VGAs) and the automatic gain control loops (AGCs) can be relaxed or even be removed if the dynamic range (DR) of the following analog-digital converter (ADC) is high enough, simplifying the analog front-end [1]. For better integration, the ADC is placed closer and closer to the antenna. Low-IF conversion with a digital channel selection filter and second-stage mixer is considered to be more cost-efficient than the traditional heterodyne receiver architecture for multi-channel applications [2]. This simplifies the RF front-end and improves the system flexibility, but calls for ADCs with larger bandwidth and higher accuracy. Therefore power-efficient ADCs with tens of MHz of bandwidth and 12–14 bit resolution are in great demand for future communication applications. Delta-Sigma modulators (DSMs) are an excellent candidate for this. For the higher bandwidth, continuous-time (CT) DSM ADCs are preferred, because of their high power efficiency [3], the inherent anti-aliasing characteristic [4] and the possibility of re-configurability [5].

This chapter presents the full designs and evaluation results of two 40 MHz-BW 12-bit CT DSM ADCs with phase-type VCO-based quantizer. In our first design [6], the CT DSM ADC is comprised of feedback DACs, a 3rd-order analog filter and a 5-bit phase-type VCO-based phase quantizer. The proposed shaped SC DAC has a reduced peak current while consuming no extra hardware or power. As a result, about 30% operational transconductance amplifier (OTA) power consumption is saved due to the relaxed slewing requirement. The front-end DAC nonlinearity is cancelled via a LUT-based digital calibration. In this way, a power-efficient 40 MHz-bandwidth modulator is obtained with 12-bit resolution and over 80 dB spurious-free dynamic range (SFDR) in a 90 nm CMOS technology. The second CT DSM is an improved design based on the first one, targeting a lower power consumption and no digital calibration [7]. Here a half-period delay but not a one-period delay scheme

X. Xing et al., *Design of Power-Efficient Highly Digital Analog-to-Digital Converters for Next-Generation Wireless Communication Systems*, Signals and Communication Technology, https://doi.org/10.1007/978-3-319-66565-8_5

is applied to compensate the modulator ELD. A capacitive local feedback without noise contribution is proposed for the NTF zero optimization. The FF OTA used in the loop filter is implemented by a current-sharing structure, leading to a considerable analog power reduction. With these techniques, about 70 dB SNR is reached with only 45 mW power, resulting in a FoM of 0.26 pJ/Step.

5.2 A 40 MHz-BW 12-Bit CT DSM with Digital Calibration and Shaped SC DAC

5.2.1 Structure of the CT DSM

5.2.1.1 System Parameters

In high-bandwidth DSMs, the signal-to-noise ratio (SNR) should be limited by the thermal noise of the analog front-end circuit [8], and the contribution of the quantization noise should be small. To incorporate some design margin, we chose to design for about 90 dB signal-to-quanzation noise ratio (SQNR) to achieve a 12-bit thermal noise limited design. In CT DSM design, the maximum OTA GBW and the minimum digital gate delay determine the maximum sampling frequency, and hence the maximum available OSR. Compared to lower-order DSMs, single-loop CT DSMs of 5th and higher order become prone to instability. Alternatively, a MASH structure could reduce the noise shaping order in each loop and improve the stability, but the mismatch between the CT integrator and the digital filter imposes another design difficulty, usually requiring digital calibration [9]. Trading off all considerations, an architecture with OSR of 12, 4th-order noise shaping and 5-bit quantization is adopted in this design for its robust performance and relaxed requirements on the building blocks.

5.2.1.2 Modulator Topology and Block Selection

The block diagram of the CT DSM is shown in Fig. 5.1: 1st-order noise shaping and 5-bit quantization are executed by a phase-type VCO-based quantizer [10]. The remaining 3rd-order noise shaping is done by the analog inverse Chebyshev filter preceding the quantizer. A non-return-to-zero (NRZ) DAC and a SC DAC delayed by one sample period are used to form the different feedback loops. To solve the modulator stability problem caused by the delayed DAC signals, an extra return-to-zero (RZ) DAC is added to form a zero-order feedback loop.

The VCO-based phase quantizer consists of a 31-stage ring VCO and 31 slices of the phase and frequency detection circuit, which is comprised of basic digital gates (flip-flop, XOR gate). The VCO converts the analog input control voltage into phase, which is indicated by the occurrence in time of the voltage transition edge.

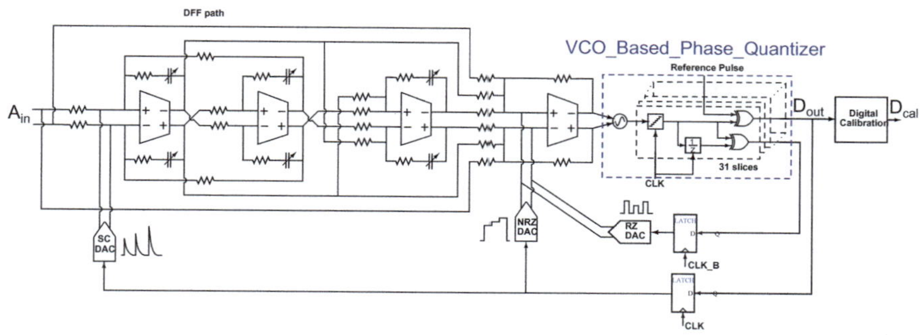

Fig. 5.1 Structure of the 40 MHz 12-bit CT DSM

The quantization error is generated when the VCO outputs are synchronized by the sampling flip-flops; the XOR gate together with the delay cell is used to differentiate the digital phase to obtain the digitized VCO frequency. The function of the second XOR gate with reference pulses is to remove the DC information of the digital phase, which corresponds to the VCO common-mode control voltage (0.6 V) and free-running frequency (one quarter of the sampling frequency). Because of the high gain of the VCO inherent integration, the input swing of the VCO is small, and as a result the effect of the nonlinearity of the voltage-to-frequency/phase relationship is greatly reduced as well. Simulation results show that the harmonic distortion of the quantizer is around −90 dBc [10], plus the attenuation provided by the loop filter. Therefore the distortion of the VCO-based quantizer is not the bottleneck for the modulator performance.

The 3rd-order analog loop filter is a feedforward structure, for its reduced integrator output swings and lower requirements for the transconductors [11]. Although the direct feedforward (DFF) path in the analog filter leads to a larger signal-transfer-function (STF) out-of-band peaking and reduced anti-aliasing filtering at higher frequencies (a deterioration of about 2 dB for both), it can reduce the integrator output swing further. This relaxes the linearity requirements for the OTA. Because of the inherent integrating operation in the VCO, at least a 1st-order inherent anti-aliasing at high frequencies is still maintained when using this DFF path. One local feedback path across the first two integrators is used to generate two finite noise-transfer-function (NTF) zeroes (at about 30 MHz) to further suppress the quantization noise near the signal bandwidth. Although Gm-C integrators are popular in high-speed CT DSM design, to achieve 12-bit linearity, both degeneration resistors and local-feedback amplifiers would be needed for the transconductors [12], at the expense of extra power. With active-RC integrators, more stable output nodes can be provided to the feedback DAC, which improves its linearity. With the optimized loop filter structure and coefficients, the amplifier GBW requirements for the second and third integrators are only two times the sample frequency. Based on these considerations, all integrators are realized as active-RC filters. The input resistors of the first integrator are about 1.5 k which leads to a thermal noise contribution of about 80 dBFS. To overcome the problem due to the large variation of the RC time constant, 4-bit binary

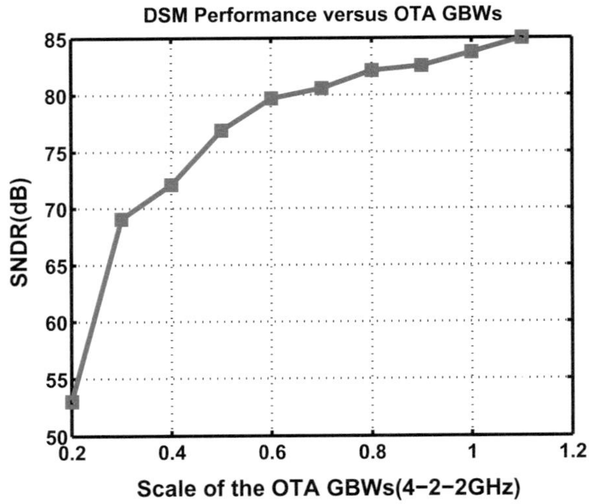

Fig. 5.2 CT DSM SNDR performance as a function of the scaled OTA GBWs (normalized to 4, 2 and 2 GHz for loop filter integrator one to three respectively)

capacitor banks are connected in parallel with the fixed-value capacitors to adjust the capacitor values in steps of 2.7% over 20% range. In CT DSM design, the OTA in the first active-RC integrator largely determines the modulator performance: insufficient gain over the whole signal bandwidth would result in a too large signal swing on the amplifier input virtual ground, increasing the effect of integrator nonlinearity and noise. A too small GBW of the OTA would induce an extra phase shift and gain attenuation, causing an increased in-band noise and distortion or even instability. Finally, a too limited voltage swing of the amplifier would lead to nonlinearity of the loop filter. Figure 5.2 shows the simulated modulator SNDR performance as a function of the scaled GBW values (from the 1st to the 3rd OTA, the GBWs are normalized to 4, 2 and 2 GHz respectively) for fixed DC gains (55 dB) of the OTAs. A GBW of 4 GHz for the first OTA is required for the modulator to keep the non-ideal OTA-induced noise/distortion below 10% of the total budget.

The performance requirements for the feedback DACs are different because of their different position in the modulator and their different inputs. Referring all non-idealities to the modulator input directly, the performance requirement for the SC DAC must be at least the same as the DSM specifications. On the other hand, the requirements of the NRZ DAC are greatly relaxed because its noise and distortion are suppressed by the 3rd-order analog filter. For the RZ DAC, in addition to the suppression provided by the loop filter, two facts further relax the required accuracy. Firstly, because of the integrating operation of the VCO, the input swing of the VCO-based quantizer is small, and so is its digital frequency output. Although the RZ DAC is 5-bit, it behaves as a lower-bit DAC most of the time (from behavioral simulation at most 6 levels are used). With most of the DAC cells switched off, both the nonlinearity and the output noise current are reduced. Secondly, in the ring VCO the inverter stages change state continuously, so the frequency output of the quantizer is intrinsically dynamic element matched, randomizing the error [13]. As they are

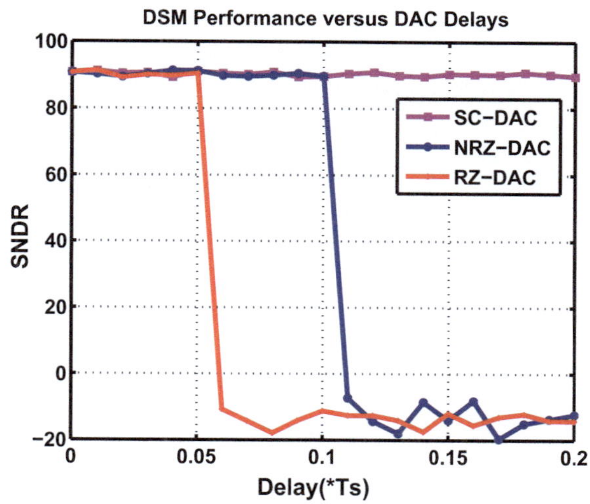

Fig. 5.3 CT DSM SNDR performance as a function of the delay for the three different feedback DACs

placed in feedback loops with different order, the speed requirements for the feedback DACs are also different. The CT DSM SNDR performance as a function of the DAC delay is presented in Fig. 5.3. It is clear that the SC DAC can tolerate as much as 20% sampling period delay, but the delay for the RZ DAC should not exceed 5% of the period. The reason is that in CT DSMs the lower-order loops are responsible for the higher-frequency feedback signal components, and for these higher-frequency signals a small delay already corresponds to a large phase shift. In this work, a shaped SC DAC with smaller peak current is proposed for the outermost DAC as it offers several advantages. The noise contribution of the SC DAC is about:

$$\overline{V_{SC\_DAC}^2} = \frac{kT}{OSR \cdot C} \tag{5.1}$$

If the total capacitor value C is chosen large enough, this noise can be ignored compared to the thermal noise of the input integrator resistor Rin:

$$\overline{V_{R_{in}}^2} = 8kTR_{in}BW \tag{5.2}$$

For comparison, if a NRZ DAC is used, then the DAC noise is about [14]:

$$\begin{aligned}\overline{V_{NRZ\_DAC}^2} &= \frac{16kTg_m}{3}R_{in}^2BW = \frac{32kTI}{3\left(V_{gs}-V_{th}\right)}R_{in}^2BW \\ &= \frac{32kTV_{ref}}{3\left(V_{gs}-V_{th}\right)}R_{in}BW\end{aligned} \tag{5.3}$$

where V_{ref} is the reference voltage of the modulator, which is 0.6 V in our design, and the gate-source over-drive voltage $V_{gs} - V_{th}$ of the current-source transistors in the NRZ DAC is usually set to 0.15 V or less in nanoscale CMOS circuit design. It is clear from (2) and (3) that in this case the NRZ DAC noise is about 5 times the noise contribution of the input resistors. Although a degeneration resistor could be used to reduce the NRZ DAC noise, a higher supply voltage is needed because of the voltage across the resistor [8]. In addition, since the output current returns to zero every period, the SC DAC is more immune to inter-symbol interference than a NRZ DAC, which is reported to be a main performance limitation in high-bandwidth CT DSMs [10]. Since it is not suppressed by any analog filter, the overall nonlinearity of the DSM is limited by that of the SC DAC, which is discussed in detail in the next subsection. With the selected architectural parameters, the DSM NTF has been synthesized by Schreier toolbox [15] and the coefficients have been optimized by behavioral-level parameter explorations.

5.2.1.3 LUT-Based Digital Calibration

As the nonlinearity of the 5-bit main DAC is not suppressed by any loop filter, it needs to be solved by other methods. If explicit DEM is applied between the quantizer and the resample flip-flop, the delay of the digital block that implements the DEM can not be larger than half a sampling period [10]. Circuit-level simulations in 90 nm CMOS show that if implemented in static logic, the delay of a 5-bit DEM block is about 740 ps, failing to meet the timing requirement. If realized in pseudo-NMOS logic, simulations show that the delay is about 470 ps, with 18.2 mW power dissipation. Considering the extra delay caused by any parasitic wiring capacitance and resistance, there is almost no design margin on the timing. Obviously also the associated large power consumption is a significant drawback.

Another solution to the nonlinearity of the main feedback DAC is to use digital calibration [16]. This concept is illustrated in Fig. 5.4. Here, for each DAC level, an accurate digital representation is stored in a digital LUT. For our case of a 5-bit DAC, 32 DAC levels must be stored. During the normal DAC operation, the corresponding calibrated output, Dcal, is then retrieved from the LUT.

Obviously the accuracy of such a calibration depends on how accurately the calibration coefficients in the LUT correspond to the actual DAC levels. To find the optimal calibration coefficients, in this work a least-mean-square (LMS)-based offline calibration approach is used.[1] A complete mathematical discussion of this technique is out of the scope of this book, but the basic idea can simply be understood. It consists of applying a known analog input signal A_{in} to the ADC. Then the calibration values in the LUT are tuned until the digital output D_{cal} matches the input signal A_{in} in a LMS sense (i.e. the mean square of $D_{cal} - A_{in}$ is minimized). In our implementation we used a spectrally pure sinusoidal input signal, which is mathematically perfectly known. For our approach it is not needed that the amplitude or the frequency are known in advance. In this way such an input stimulus signal can

[1]This technique was developed by Pieter Rombouts and Maarten De Bock. It was applied in this design as part of a collaboration within a FWO project.

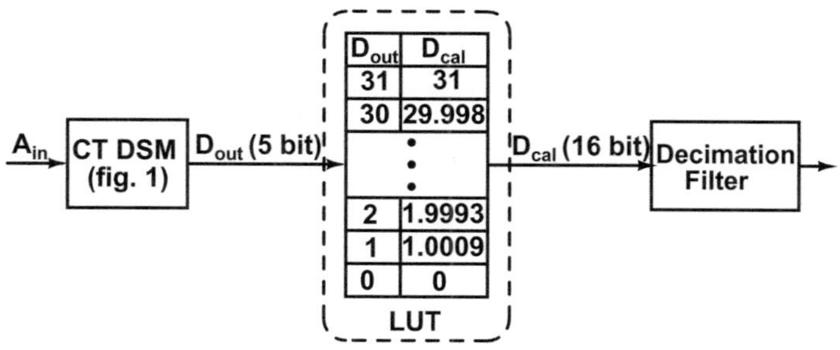

Fig. 5.4 Block diagram of the digital calibration of the DAC. The values in the look-up table are arbitrary values, for illustrative purpose

readily be obtained in practice through bandpass filtering. Clearly this approach is very well suited for prototype debugging and testing. Moreover, the implementation of this LMS algorithm we used requires only 1024 baseband samples and hence is also a viable approach for factory-level calibration.

5.2.2 Delta-Sigma Modulator Building Blocks Design

5.2.2.1 4-Stage Feedforward Compensated OTA

For OTAs with traditional structures, there are different bottlenecks to realize the specification of 55 dB gain and 4 GHz GBW. With a single-stage amplifier, it is difficult to reach a gain of over 30 dB in 90 nm CMOS. In cascoded and folded-cascoded OTAs, either the supply voltage or the output swing is limited. For a Miller-compensated topology, too much power is required to push all non-dominant poles far beyond the GBW. Therefore, a 4-stage full-feedforward compensated OTA is proposed in this design to achieve a high gain, a high GBW and a large signal swing. The block diagram is shown in Fig. 5.5, and all transconductors are realized by NMOS differential pairs, with PMOS current sources as loads. The 4 cascaded stages generate 4 poles. The stability is guaranteed by introducing 3 left-half-plane (LHP) zeros to cancel or compensate 3 poles. Here each feedforward path generates one LHP zero [17].

$$\frac{V_{out}}{V_{in}} = \frac{G_{m1}R_1G_{m2}R_2G_{m3}R_3G_{m4} + sG_{mf3}C_1R_1}{1 + sC_1R_1} \frac{R_4}{1 + sC_LR_4} \tag{5.4}$$

$$P_1 = -\frac{1}{C_1R_1} \quad P_4 = -\frac{1}{C_LR_4} \quad Z_3 = -\frac{G_{m1}G_{m2}R_2G_{m3}R_3G_{m4}}{G_{mf3}C_1} \tag{5.5}$$

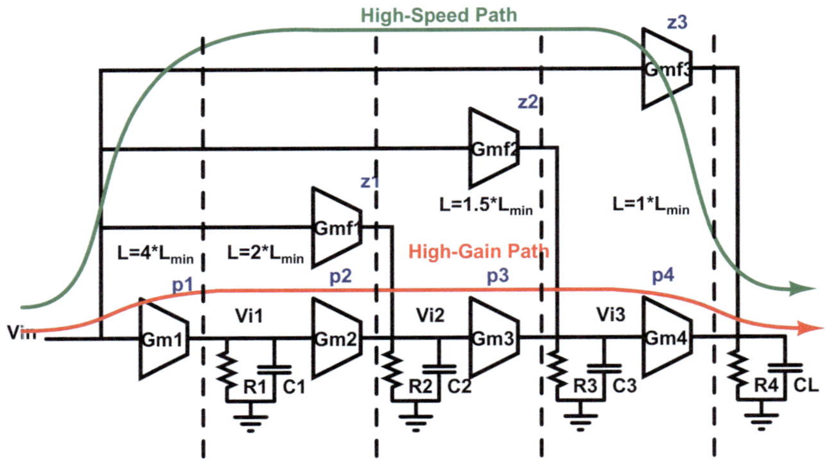

Fig. 5.5 Block diagram of the 4-stage full-feedforward compensated operational transconductance amplifier

In our design, because of the large output resistance R_1 and the small output resistance R_4, P_1 is the dominant pole (about 7.7 MHz) and P_4 is the non-dominant pole (about 600 MHz); Z_3 (at about 200 MHz) is placed between P_1 and P_4 to increase the GBW and the phase margin (PM). For the 1st- and 2nd-stage cascaded amplifiers, the output signal swings are small, so PMOS cascode transistors are added to increase the gain and to reduce the output parasitic capacitors. The transistor gate lengths of the 4 stages are 4, 2, 1.5 and 1 the minimum gate lengths respectively. For the internal high-impedance nodes, capacitors are added intentionally to outweigh the transistor parasitic capacitors, leading to controllable poles and zeroes. A Miller-compensated common-mode feedback loop (CMFB) with about 500 MHz GBW is arranged for every stage to get a stable CM voltage. In CMOS implementation, because of the technology parameter variations, the above cancellation of the pole-zero pairs may not be exact. In Fig. 5.6, the modulator SNDR performance with OTA doublets is plotted. In these simulations, three fixed poles (30/200/600 MHz) and three zeroes with variable positions are considered in all OTAs. It is shown that 90% normalized doublet frequency variation can be tolerated by the proposed design, and there is no SNDR performance penalty if the doublet relative position is within this range. In Fig. 5.7, the frequency response of the first OTA is presented. Also the internal gains of the three cascaded stages are shown. The proposed OTA reaches 56 dB DC gain, 3.9 GHz GBW and 82° phase margin with a 1.3 pF load capacitor, dissipating 14.4 mW power. Because of the large power consumption, the input noise (from 1 kHz to 40 MHz) is only 16 μV. As their non-idealities are shaped by the preceding integrators, the second- and third-integrator amplifiers have the same structure, but with about 40% reduced size and power.

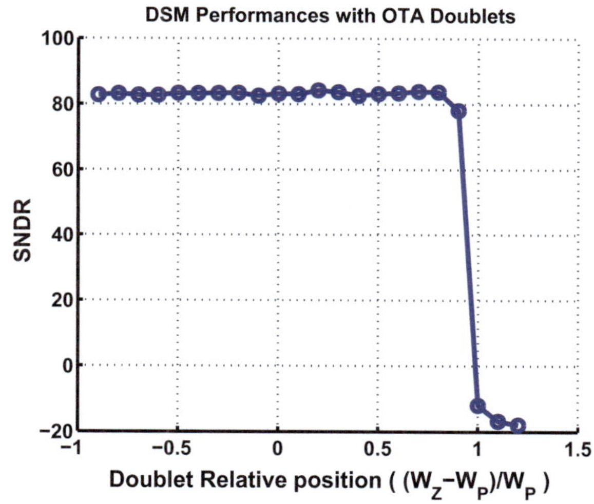

Fig. 5.6 SNDR performance of the CT DSM as a function of the relative position of the poles and zeros in the doublets

5.2.2.2 Shaped Switched-Capacitor DAC with Reduced Peak Current

For the SC DAC design, three parameters are critical if applied in a CT DSM: the peak current I_{peak}, the tail current I_{tail} and the delivered charge Q. The peak current determines the slew-rate requirement in the first OTA; the tail current decides the modulator clock jitter sensitivity, and the delivered charge sets the feedback coefficient:

$$I_{peak} = \frac{V_{ref}}{R_{on}} \quad I_{tail} = \frac{V_{ref}}{R_{on}} exp\left(\frac{-0.5T}{\tau}\right) \tag{5.6}$$

$$Q = V_{ref}C\left[1 - exp\left(\frac{-0.5T}{\tau}\right)\right] \quad \tau = R_{on}C_u \tag{5.7}$$

where V_{ref} is the reference voltage, R_{on} is the on-resistance of the switches, C_u is the capacitor of the SC DAC cell and τ is the discharge time constant. To relax the OTA slew rate, one can reduce the peak discharge current, by increasing the on-resistance R_{on} or by reducing the reference voltage V_{ref}. However, to keep the feedback coefficient (the charge Q) the same, a larger discharge time constant is needed, which will result in a larger discharge tail current and impair the DSM clock jitter performance. We solve this problem using a shaped SC DAC cell.

The shaped SC DAC cell and its non-overlapping driving circuit are shown in Fig. 5.8: the capacitors are implemented as metal-insulator-metal (MIM) capacitors and the switches are realized as transmission gates. The capacitor of each SC DAC cell C_u is 200 fF, so the SC DAC noise is about −95 dBFS. In the charging phase, the two capacitors are charged to the positive and negative reference voltages respectively. In the discharging phase, the charges are injected onto the OTA virtual grounds and

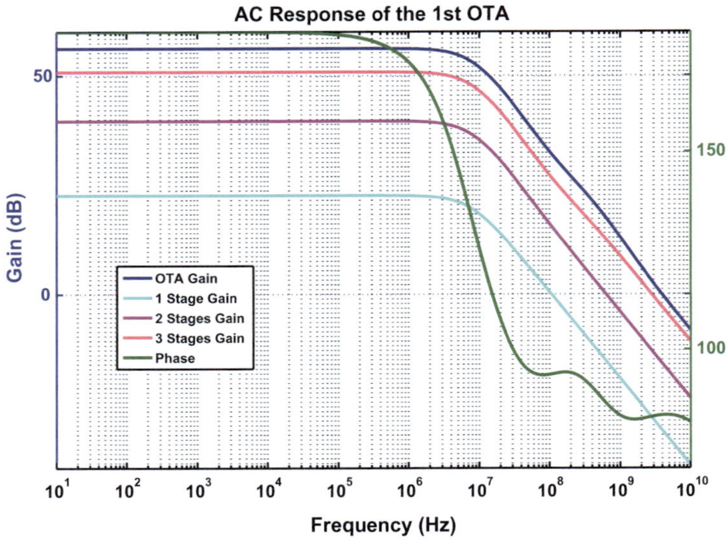

Fig. 5.7 AC simulation result of (internal) gain of the operational tranconductance amplifier used in the first integrator

integrated on the first integration capacitors. The discharge polarity is determined by the DAC input, and the shape of the discharge current depends on the rising/falling edge slope of the DAC driving signal, the reference voltage V_{ref}, the on-resistance of the switches R_{on} and the discharge time constants. Compared with a traditional SC DAC, our proposed shaped SC DAC introduces the driving signal slope to shape the DAC discharge current and breaks down the trade-off between the OTA slew rate requirement and the modulator clock jitter sensitivity. It is observed that if the rise time of the discharge current is enlarged, the peak current is reduced, and so is the slew rate requirement in the first OTA.

In our design, the rising edge of the DAC driving signal is low-pass filtered, which can be realized by adjusting the transistor drive capability of the preceding gates. As shown in Fig. 5.8, in the driving circuit for the transmission gates NMOS part, the PMOS transistor width is reduced to realize an extended driving signal rising edge; in the driving gate for the PMOS transmission transistor, the NMOS transistor size is scaled down for a low-passed filtered falling edge. As a result, the rising edge of the DAC discharge current gets one extra time constant, resulting in an extended rise time (about 72 ps) and a reduction of about half of the current peak value, as shown in Fig. 5.9. Since this effect only occurs during the switching-on of the switches, the tail of the DAC discharge current is not affected by this shaping, and the current amplitude at the end of the discharging phase is almost the same (increased only 8% compared to the ideal discharge current). With this shaped SC DAC current pulse, the slew-rate requirement of the first OTA can greatly be relaxed, reducing its power consumption by about 30%. Because of the SC DAC incompletely settling in CT

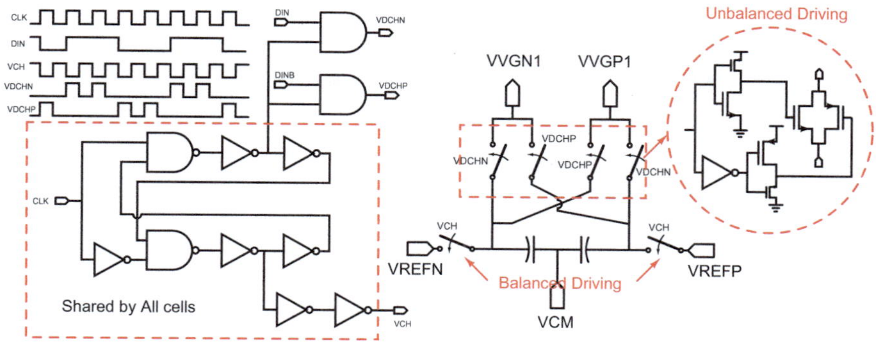

Fig. 5.8 Circuit implementation of the shaped SC DAC (*right*) and its driving circuit (*left*)

Fig. 5.9 The output current pulse of the shaped SC DAC, compared to the traditional pulse shape, as used in our design

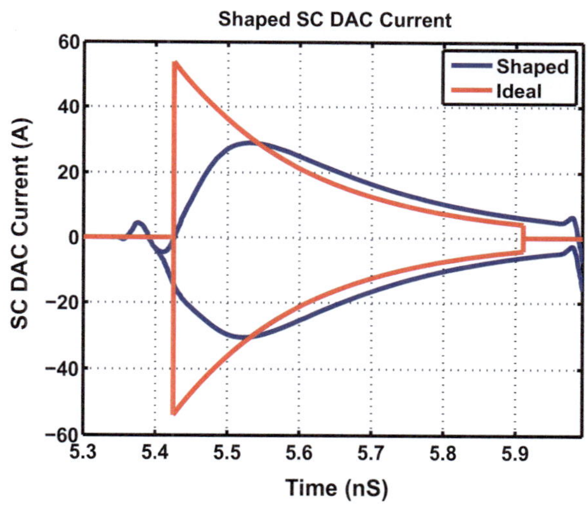

DSMs, the matching of the SC DAC cells does not only depend on the matching of the capacitors, but also on that of the on-resistance of the transmission gates (assuming no mismatch on the driving gates):

$$\sigma^2\left(\frac{\Delta Q}{Q}\right) = \left(\frac{mTexp(-mT/\tau)/\tau}{1-exp(-mT/\tau)}\right)^2 \sigma^2\left(\frac{\Delta R_{on}}{R_{on}}\right) + \left(\frac{1-exp(-mT/\tau)-mTexp(-mT/\tau)/\tau}{1-exp(-mT/\tau)}\right)^2 \sigma^2\left(\frac{\Delta C}{C}\right) \quad (5.8)$$

Here mT is the fall time of the shaped SC DAC discharge current, which is about 410 ps in our design. With the component sizing and mismatch information, the mismatch of the redistributed charge is about 1%, which could be solved by the digital calibration described in the last subsection.

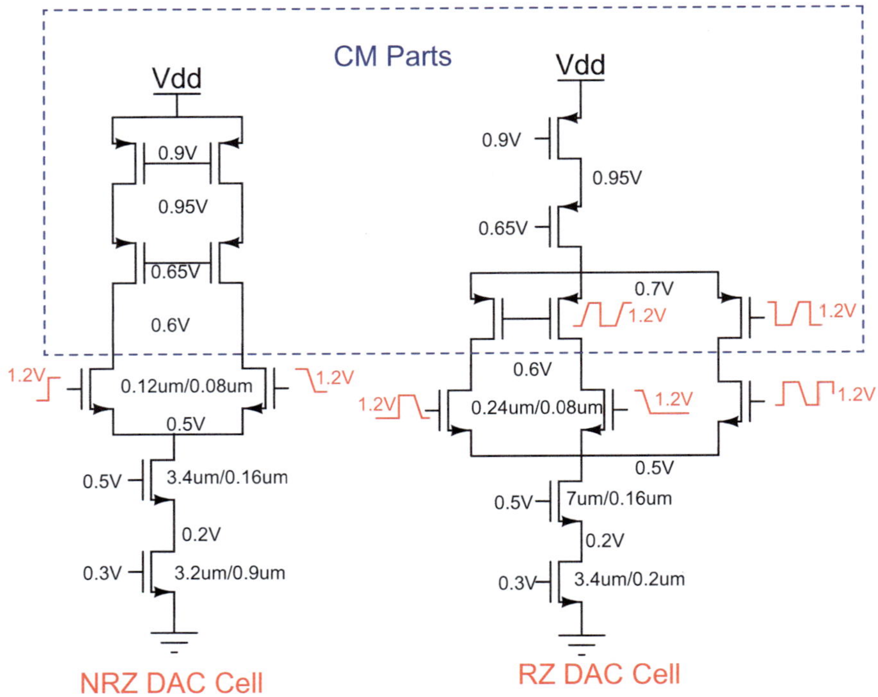

Fig. 5.10 Schematic of the current-steering DAC cells

5.2.2.3 Current-Steering DACs

The schematics of the NRZ and RZ DAC cells and their biasing voltages are shown in Fig. 5.10: cascode transistors with about 8× gain are used to shield the current-source transistors from the noisy switching transistors. Overlapping drive signals (high crossing for the NMOS part and low crossing for the PMOS part) are applied to minimize the glitch energy [18]. Because full swing is used for the driving signals, the sizes of the switch transistors are small, leading to a high-speed operation.

5.2.2.4 Adder

Both passive and active adders could be used to convert the CS DAC currents to voltages and sum them together with the integrator output voltages. However, compared with an active adder, a passive adder can only provide attenuation, especially when there are several input branches. The VCO input capacitor is around 235 fF. If a passive adder would be applied, small resistors are needed to push the poles formed by the adder resistors and this capacitor to a high frequency to avoid an extra phase shift, which means a large AC power. For these reasons, an active adder with about

7 mW power is applied in our design. One more advantage of the active adder is the more stable outputs for the DACs, resulting in a better linearity of these DACs.

5.2.3 *Measurement Setup and Experimental Results*

5.2.3.1 Measurement Setup

The 40 MHz 12-bit CT DSM has been manufactured in UMC 90 nm standard-performance (SP) CMOS. The chip photo is shown in Fig. 5.11. The area of the core circuit is about 0.28 mm^2. The 31 thermometer bits are coded to 5-bit binary outputs by a Wallace tree adder (the U2B block on the chip photo) and fed to low-voltage differential signaling (LVDS) buffers to drive the measurement instruments. The measurement setup for the designed CT DSM is shown in Fig. 5.12: a ceramic substrate has been designed to mount the die and to provide electrical connections with the instruments. All power supplies, bias voltages/currents and control bits are generated by a FR4 printed circuit board (PCB). All supply voltages and bias nodes are decoupled by capacitors on the ceramic substrate and the PCB. Passive bandpass filters are used to suppress harmonics and noise of the analog input signals from the AWGs. For the two-tone test a power combiner combines two filtered sine waveforms without introducing intermodulation distortion (IMD). A balun converts the single-ended analog waveforms into differential signals. The clock signal is generated by a pulse generator with low jitter (typical value of 1 ps rms). The on-chip LVDS digital output bits are captured by a logic analyser and then analysed by Matlab code. In our setup the digital calibration described previously has been implemented in Matlab and performed in an offline cycle after which the calibration coefficients are stored in the memory. The calibration algorithm used 16 K samples (corresponding to 1365 baseband samples). The execution of the calibration script on a low-cost desktop PC was nearly instantaneous (<0.2 s). Based on the observed calibration values, the mismatch of the DAC unit elements was found to be in the order of 1%, most of which is from the on-resistance of the transmission gates.

5.2.3.2 Result and Discussion

On the left of Fig. 5.13, the measured power spectral density is plotted for the uncalibrated digital outputs of -4.5 dB single-tone inputs with frequencies of 938 kHz and 7.5 MHz. On the right side, the corresponding calibrated results are shown. The DAC mismatch distribution used for calibration is extracted from the measurement with a 938 kHz input. When a 938 kHz input is applied, the SNDR without and with calibration is 63.5 dB and 70.8 dB, respectively. As seen in the plot, after calibration, all harmonics are under the noise floor. In the 7.5 MHz input measurement, by applying the digital calibration, the SNDR is improved from 63.1 to 70.3 dB. At this input frequency, part of the 2nd harmonic remains. For both frequencies, the SNDR

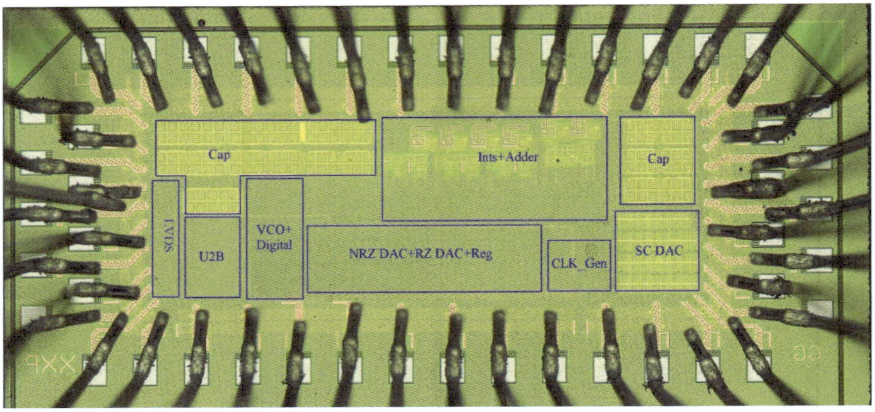

Fig. 5.11 Die photo of the 40 MHz 12-bit CT DSM

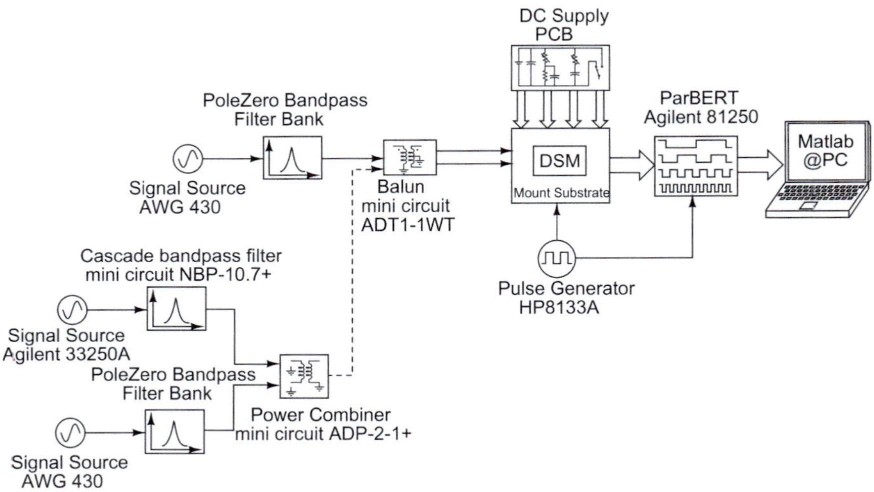

Fig. 5.12 Measurement setup for the 40 MHz 12-bit CT DSM

is improved by more than 7 dB, validating the effectiveness of the digital calibration. The out-of-band peak in the measured spectrum of Fig. 5.13 is a result of the changed NTF, and can be adjusted by means of the capacitor banks in the integrators and the DAC biases. The measured modulator SNR/SNDR for increasing analog input powers is shown in Fig. 5.14. For the 938 kHz sine input, the peak SNR is 73.6 dB with a −2 dB analog input, and the peak SNDR is 72.9 dB when the input is −2.5 dB. The peak SNR/SNDR is 71.7/70.8 dB respectively when a 7.5 MHz −3/−3.5 dB analog input is applied. The peak SFDR for the two measurements is 84.2 and 80.6 dB, respectively, which is very large for a 40 MHz ADC.

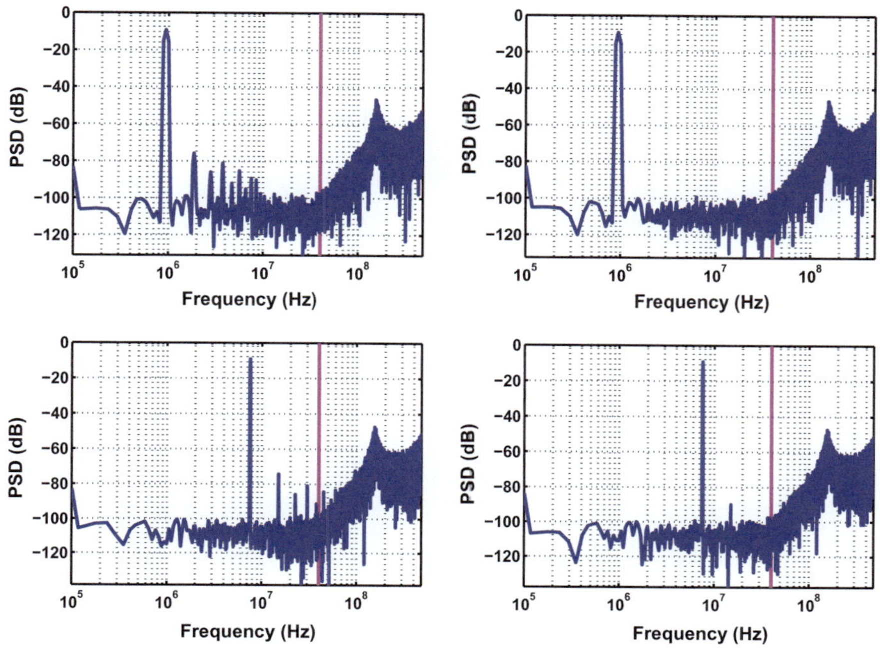

Fig. 5.13 Measured spectra with (*right*) and without (*left*) digital calibration for 938 kHz and 7.5 MHz input signals

In Fig. 5.15, the spectrum of a −10.5 dB two-tone input (10 and 10.5 MHz) IMD test is shown. Because of the digital calibration, the 3rd-order IMD is almost buried in the noise floor, and the 2nd-order IMD is about −78 dB. The power dissipation of the core circuit is 69.6 mW, of which about 45 mW is consumed by analog blocks, including the integrators, adder and DACs. At low input frequencies (1 MHz) the circuit achieves a FoM of 0.31 pJ/Step (Power/(2BW*2^{ENOB})) at the worst-case input frequency. The overall performance of the implemented CT DSM is summarized in Table 5.1, and will be compared to the state-of-the-art works at the end of Chap. 8, after having presented the final ADC design.

5.3 A 40 MHz-BW 12-Bit CT DSM with Capacitive Local Feedback and Current-Sharing OTA

In the design of the previous section, the foreground digital calibration can not trace the temperature and other (drift, ageing, etc.) variations. To reduce the modulator power consumption and apply the DSM without digital calibration, an improved design of the 40-MHz 12-bit CT DSM is done and presented in this section.

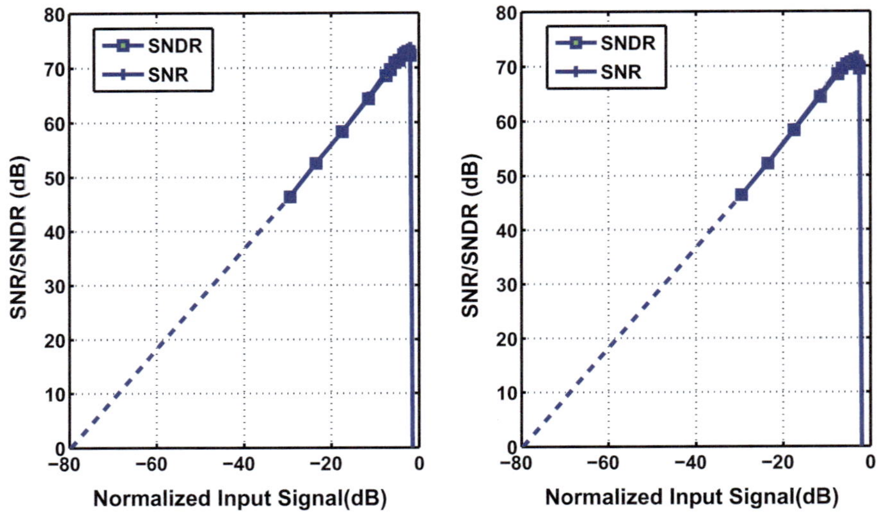

Fig. 5.14 SNR/SNDR versus input amplitude for 938 kHz input signals (*left*) and 7.5 MHz input signals (*right*)

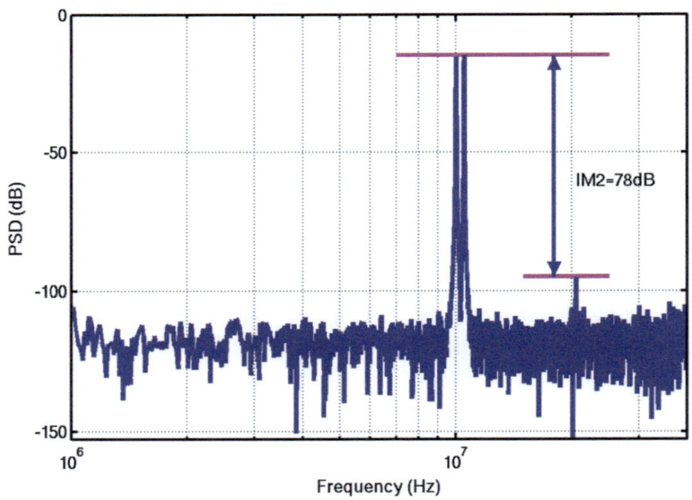

Fig. 5.15 Output spectrum for a −10.5 dB two-tone test with 10 and 10.5 MHz frequencies

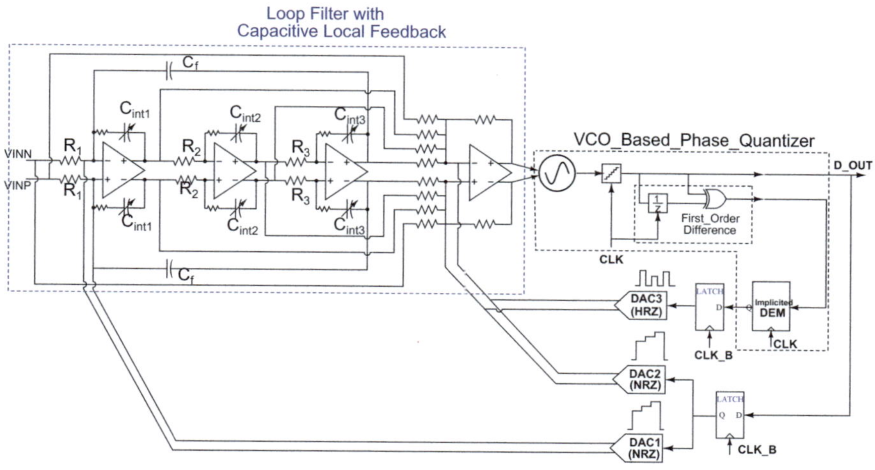

Fig. 5.16 Block diagram of the 40 MHz 12-bit CT DSM

5.3.1 *System Design of the 40 MHz 12-Bit CT DSM*

5.3.1.1 System Parameter Considerations

The noise budgets of wideband DSMs are typically designed such that the thermal noise is dominant. To ensure this, the signal-to-quantization-noise ratio (SQNR) is typically chosen to be 10–15 dB higher than the target specification [8]. In theory, many combinations of OSR, number of quantization bits and noise shaping order can reach the required SQNR. The sampling frequency is limited by the speed of the digital gates and the OTA GBW. With a lower clock frequency, the modulator is less sensitive to clock jitter. This is important from the perspective of the application and measurements. For stability reasons, the order of noise shaping of a single-loop low-pass DSM should not be over 5th [5]. Multi-stage noise shaping (MASH) on the other hand can achieve a higher order of noise shaping and still maintain the modulator stability. But the matching between the analog circuits and the digital noise cancellation filters is a problem, especially for CT DSMs. A higher number of quantization bits means less quantization noise, which improves the modulator stability and the clock jitter sensitivity if NRZ feedback DACs are used. Normally it should not exceed 5 bits because of the resulting digital complexity and power consumption. Within these boundary constraints, the architecture with 12× OSR, 5-bit quantization and 4th-order noise shaping is chosen for its robust operation and the relatively moderate requirements on the building blocks.

Table 5.1 Measurement summary of the CT DSM ADC with digital calibration and shaped SC DAC

Specification	Value
Signal bandwidth	40 MHz
Sampling frequency	0.96 GHz
Peak SNR	71.7 dB
Peak SNDR	70.8 dB
Power	69.6 mW
FoM_1	0.31 pJ/Step
FOM_2	158.4 dB
Core area	0.28 mm^2
Technology	UMC 90 nm SP

5.3.1.2 Architecture of the 40 MHz 12-Bit CT DSM

The architecture of the CT DSM is shown in Fig. 5.16. To make use of the digital circuit potential in accuracy and power efficiency in advanced CMOS technologies, the 1st-order noise shaping and 5-bit quantization are implemented by a power-efficient VCO-based quantizer [10], which is composed of a 31-stage VCO and 31 slices of frequency and phase detectors. The remaining 3rd-order noise shaping is done by an analog filter based on active-RC integrators, which have better linearity than Gm-C integrators. In the feedforward loop filter, a noise-free capacitive local feedback is used to generate a pair of complex NTF zeros to suppress the quantization noise further. A direct-feedforward (DFF) path from the modulator input to the quantizer input is applied to keep the output signal swing of each integrator within 100 mV. This deteriorates the anti-aliasing filtering at high frequencies, but due to the integration performed by the VCO, 1st-order anti-aliasing filtering is still maintained in this CT DSM. The feedback DAC signals are delayed by half a clock period to absorb the excess loop delay from the quantizer to the resample flip-flop (about 210 ps without parasitics). An extra RZ DAC is added to feed back the VCO digital frequency. This is needed to compensate for the explicit loop delay of half a clock cycle. Compared to using a full clock cycle delay in the feedback path, in our configuration more phase lag can be tolerated in the feedback DACs and the loop filter. As a result, a smaller OTA GBW is required to reach the same performance. The half-period-delayed architecture also makes the modulator more robust to other nonidealities of the analog parts. For instance, the 3rd DAC delay (6% of the period for the one-period-delayed architecture versus 10% for the half-period-delayed one) and the integrator time constant negative variation (−10% for the one-period-delayed architecture versus −13% for the half-period-delayed one) are relaxed. As no analog filter is preceding the 1st DAC, its matching should not be worse than the DSM linearity requirement. For the 2nd DAC, as its nonidealities are suppressed by the 3rd-order analog filter, 5% mismatch can be tolerated for 12-bit linearity. The intrinsic dynamic element matching (DEM) of the VCO-based quantizer and the low signal swing further relax the accuracy requirements for the 3rd DAC. Behavioral simulations show that its nonlinearity has almost no effect on the whole modulator performance.

5.3.2 Circuit Design of the DSM Building Blocks

5.3.2.1 Capacitive Local Feedback

The schematic of the loop filter with capacitive local feedback is shown in blue dashed box in Fig. 5.16. The 1st OTA, the feedback capacitor C_f and the 1st integration capacitor C_{int1} form a capacitive attenuator with a gain of C_f/C_{int1}. The output of the 3rd integrator is attenuated by this gain and fed into the 2nd integrator to build a resonator, generating two complex NTF zeros. Compared with the traditional local feedback implementation, the noise-free capacitor C_f is proposed to replace a noisy resistor. Because of the small feedback coefficient, the value of the feedback capacitor C_f is small compared to the integration capacitor C_{int3}, requiring very limited area and power consumption.

5.3.2.2 Current-Sharing Feedforward OTA

In DSMs with high-order noise shaping, the integrator OTAs are the most power-hungry blocks. For example, they take 52% of the total power consumption in [10]. The thermal noise contribution of the OTAs is a small fraction of the noise budget. So in order to implement low-power high-order DSMs, it is highly needed to design power-efficient OTAs. In feedforward compensated OTAs [6], the amplification is split into a high-gain path and a high-speed path to avoid the trade-off conflict between gain and bandwidth. Since all poles are within the GBW and no compensation capacitor is used, more power can be saved compared to the Miller OTA. The schematic of the 4-stage current-sharing feedforward (CS FF) OTA is shown in Fig. 5.17. At the outputs of the 4 cascade stages, 4 poles are generated. To ensure the OTA stability, 3 FF paths are introduced from the input to the different stage outputs to generate 3 left-half-plane zeros that cancel or compensate the poles. Compared to previously published FF-compensated OTAs, the FF transconductances (Gms) are implemented by PMOS differential pairs, which can share their bias currents with the cascade Gms.

The CS technique improves the OTA design in three aspects. Because of the current sharing between the FF Gms and the cascade Gms, less current is consumed to reach the same GBW compared to the traditional FF OTA. The FF Gms and cascade Gms load each other and less current is used in the last three branches. As a result, a larger small-signal impedance is obtained in each stage and the amplifier gain is improved by more than 10 dB. Finally, differential pairs but not simple PMOS current sources connect the outputs and the power supply, so the isolations between these nodes are better. Besides the higher gain, the power supply rejection ratio (PSRR) is improved by 13 dB, making it more robust to noise coupled from the digital part. In the 3rd stage the transconductance difference between the FF Gm and the cascade Gm is too large, so extra NMOS bias transistors are added to provide more bias current for the FF Gm, avoiding pushing the PMOS transistors into the subthreshold region. Thanks to the DSM feedforward architecture and the VCO integration, the output voltage swing of each OTA is small enough (<100 mV) to guarantee having no nonlinearity penalty due to the stacked transistors. The SPICE simulation results in Fig. 5.18 show that, with 1.3 pF load and 8 mW power consumption from a 1.2 V supply, the CS FF OTA reaches 66 dB DC gain and 3.47 GHz GBW. The phase margin and input-referred noise are 66.2° and 18 Vrms (from 1 kHz to 40 MHz) respectively. Since non-idealities are shaped by the preceding stages, the size and power of the other OTAs are scaled by 50%.

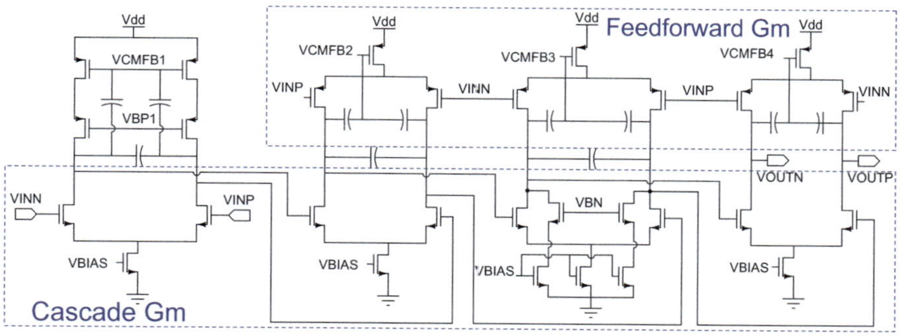

Fig. 5.17 Schematic of the current-sharing feedforward OTA

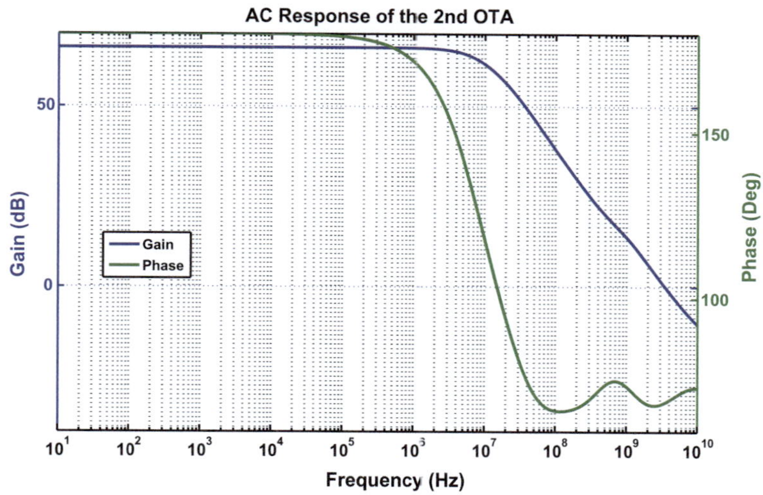

Fig. 5.18 AC simulation results of the current-sharing feedforward OTA

5.3.2.3 Feedback DACs

In this work, a current-mode NRZ DAC [18] consisting of an array of cascode current sources is used as the main feedback DAC. For the current source transistors, the overdrive voltage is set to a large value to reduce the noise current. A large area ($18\,\mu m^2$) is used to achieve a good intrinsic matching. In the layout design, double-centroid layout and tree-structure connections are adopted to cancel the mismatch further. For the cascode transistors, a gate length of two times the minimum length is used to reach a higher output impedance and an effective glitch attenuation from the noisy switching node to the quiet current-source transistor. Low-swing (400 mV) high-crossing driving signals are used to obtain a small glitch energy; they also put the switch transistors in saturation, increasing the DAC impedance further. Simulations show that the input-referred DAC noise (noise of the input resistors included) is

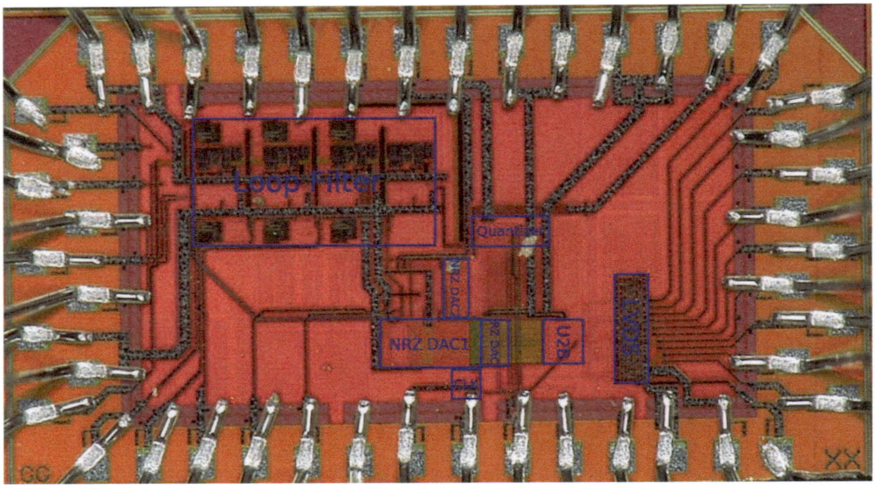

Fig. 5.19 Die photo of the low-power 40 MHz 12-bit CT DSM

about 72 Vrms, and the output impedance and output parasitic device capacitor of the whole DAC are about 21 kΩ and 66 fF respectively. In the 2nd and 3rd DAC, the sizes of the current-source transistors are scaled largely due to their relaxed impact on the modulator noise and matching. Full-swing driving signals are adopted for high-speed operation and small output parasitic capacitors.

5.3.3 *Measurement Results and Discussions*

This modulator has been fabricated in UMC 90 nm SP CMOS, with a 0.14 mm^2 core area. The micro photo of the die is shown in Fig. 5.19. The core is surrounded by MOS decoupling capacitors. In Fig. 5.20 the power spectral density (PSD) of single-tone measurements is presented. Because of the residual mismatch in the main DAC, some harmonic distortions appear and limit the performance. The peak SFDR values of the 938 kHz and 7.5 MHz measurements are 78.3 dB and 77.2 dB respectively. Figure 5.21 shows the relationship between the input power and the SNR/SNDR. The peak SNDR values are 69.7/68.4 dB and the DRs are 74.4/71.9 dB for a 938 kHz and a 7.5 MHz input frequency. A two-tone test with a 10 MHz and a 10.5 MHz input is given in Fig. 5.22. A peak IM3 of 75.1 dB is obtained with −9.5 dBFS input. The power consumption of the CT DSM is 45 mW, 62% of which is used for the OTAs and the DACs. The corresponding FoM is 0.22 and 0.26pJ/Step (Power/(2BW*2^{ENOB}))

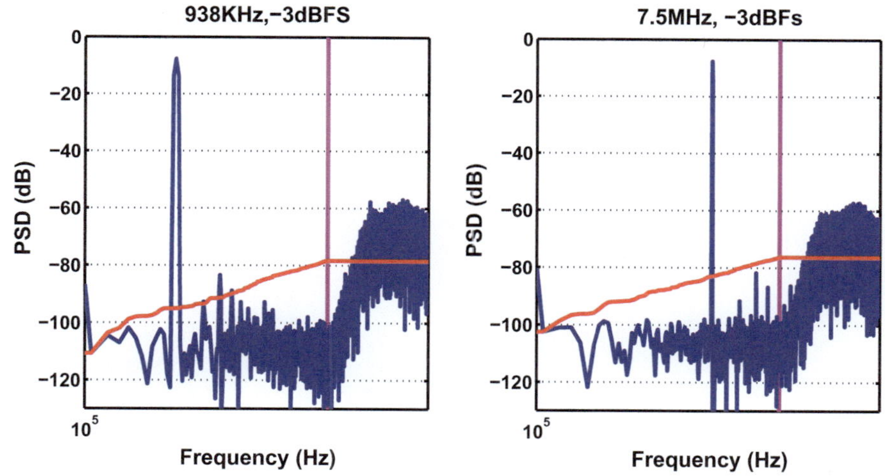

Fig. 5.20 Measured spectra for a −3 dBFS 938 kHz (*left*) and a 7.5 MHz (*right*) single-tone measurement

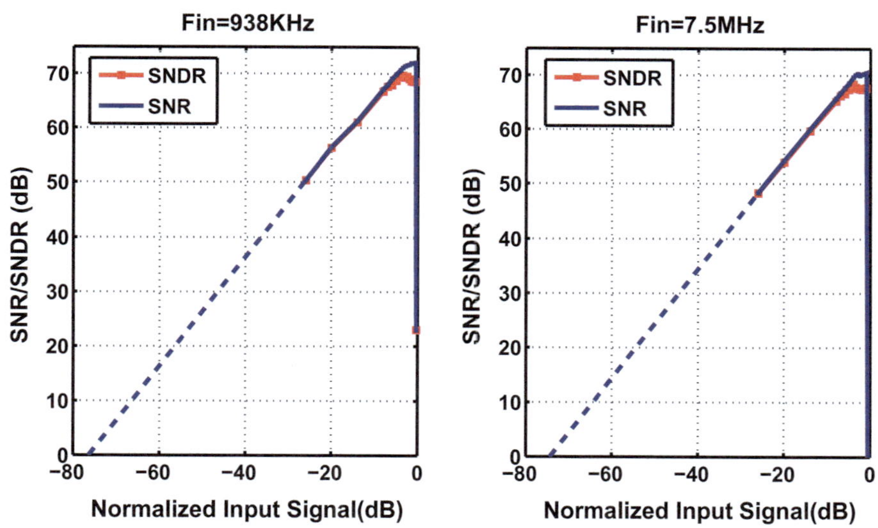

Fig. 5.21 Measured SNR/SNDR versus input amplitude for a 938 kHz input signal (*left*) and a 7.5 MHz input signal (*right*)

respectively for the two input frequencies. The CT DSM performance is summarized in Table 5.2. At the end of Chap. 8, after having presented the final ADC design, this work is compared to the published state-of-the-art DSM ADCs.

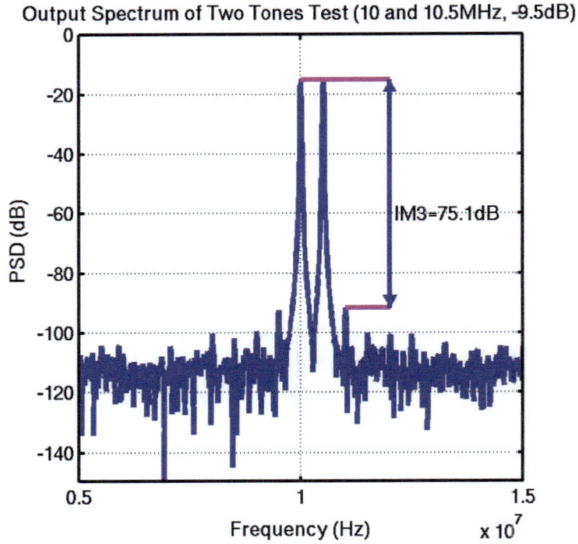

Fig. 5.22 Spectrum of two-tone measurement (−9.5 dBFS, 10 MHz and 10.5 MHz input signals)

Table 5.2 Measurement summary of the CT DSM ADC with capacitive local feedback

Specification	Value
Signal bandwidth	40 MHz
Sampling frequency	0.96 GHz
Peak SNR	70.5 dB
Peak SNDR	68.4 dB
Power	45 mW
FoM_1	0.26 pJ/Step
FoM_2	157.9 dB
Core area	0.14 mm^2
Technology	UMC 90 nm SP

5.4 Conclusions

In this chapter, the full design procedures and measurements results of two 40 MHz-bandwidth 12 bit CT DSM ADCs with phase-type VCO-based quantizer have been presented. In the first design, the SC DAC peak current is reduced by a cost-efficient shaping technology, and the DAC static and dynamic mismatches are cancelled by a digital LUT-based calibration. 4-stage FF OTAs have been designed to meet the stringent system requirements caused by the large bandwidth. The power consumption of the second CT DSM is optimized by using a half period delay architecture and a current-sharing FF OTA. A noise-free capacitive local feedback is proposed to generate the non-zero NTF zeros.

The measurement results of the two prototypes in 90 nm CMOS show that with a 960 MHz sampling frequency the first CT DSM achieves 72.9 dB SNDR and 84.2 dB SFDR with 69.6 mW power consumption, while the peak SNDR of the second CT DSM is 68.5 dB with only 45 mW power. Their FoM is 0.31 and 0.26 pJ/Step, respectively, comparing favourably to other published designs with a high signal bandwidth. The second design has an improved energy efficiency, but it is still higher than other DSM designs with lower bandwidth. The next chapter will therefore present a third design that improves significantly the energy efficiency.

References

1. P.G.R. Silva, L.J. Breems, K.A.A. Makinwa, R. Roovers, J.H. Huijsing, An IF-to-baseband Sigma-Delta modulator for AM/FM/IBOC radio receivers with a 118 dB dynamic range. IEEE J. Solid-State Circuits **42**(5), 1076–1089 (2007)
2. L.J. Breems, R. Rutten, R.H.M. van Veldhoven, G. van der Weide, A 56-mW continuous-time quadrature cascaded modulator with 77-dB DR in a near zero-IF 20 MHz band. IEEE J. Solid-State Circuits **42**(12), 2696–2705 (2007)
3. G. Mitteregger, C. Ebner, S. Mechnig, T. Blon, C. Holuigue, E. Romani, A 20-mW 640-MHz CMOS continuous-time ADC with 20-MHz signal bandwidth, 80-dB dynamic range and 12-bit ENOB. IEEE J. Solid-State Circuits **41**(12), 2641–2649 (2006)
4. F. Gerfers, M. Ortmanns, *Continuous-Time Sigma-Delta A/D Conversion: Fundamentals, Performance Limits and Robust Implementations* (Springer, Berlin, 2006)
5. Y. Ke, Design of reconfigurable A/D converter for 4G radios. Ph.D. thesis, KU Leuven, Heverlee, Belgium, 2011
6. X. Xing, M. De Bock, P. Rombouts, G. Gielen, A 40 MHz 12 bit 84.2 dB-SFDR continuous-time Delta-Sigma modulator in 90 nm CMOS, in *2011 IEEE Asian Solid State Circuits Conference (A-SSCC)* (2011), pp. 249–252
7. X. Xing, G. Gielen, 40 MHz-BW continuous-time $\Delta\Sigma$ modulator with capacitive local feedback and current-sharing OTA. Electron. Lett. **49**(9), 585–586 (2013)
8. M. Bolatkale, L.J. Breems, R. Rutten, K.A.A. Makinwa, A 4 GHz CT ADC with 70 dB DR and −74 dBFS THD in 125 MHz BW, in *2011 IEEE International Solid-State Circuits Conference Digest of Technical Papers (ISSCC)* (2011), pp. 470–472
9. G. Cauwenberghs, G.C. Temes, Adaptive digital correction of analog errors in MASH ADCs I. Off-line and blind on-line calibration. IEEE Trans. Circuits Syst. II Analog Digit. Sig. Process. **47**(7), 621–628 (2000)
10. M. Park, M.H. Perrott, A 78 dB SNDR 87 mW 20 MHz bandwidth continuous-time $\Delta\Sigma$ ADC with VCO-based integrator and quantizer implemented in 0.13 μm CMOS. IEEE J. Solid-State Circuits **44**(12), 3344–3358 (2009)
11. J. Silva, U. Moon, J. Steensgaard, G.C. Temes, Wideband low-distortion Delta-Sigma ADC topology. Electron. Lett. **37**(12), 737–738 (2001)
12. Z. Chang, D. Haspeslagh, J. Verfaillie, A highly linear CMOS Gm-C bandpass filter with on-chip frequency tuning. IEEE J. Solid-State Circuits **32**(3), 388–397 (1997)
13. M.Z. Straayer, M.H. Perrott, A 12-bits, 10-MHz bandwidth, continuous-time Delta-Sigma ADC with a 5-bits, 950-MS/s VCO-based quantizer. IEEE J. Solid-State Circuits **43**(4), 805–814 (2008)
14. Z. Li, T.S. Fiez, A 14 bit continuous-time Delta-Sigma A/D modulator with 2.5 MHz signal bandwidth. IEEE J. Solid-State Circuits **42**(9), 1873–1883 (2007)
15. R. Schreier, G.C. Temes, *Understanding Delta-Sigma Data Converters* (Wiley, New York, 2005)

16. M. De Bock, X. Xing, L. Weyten, G. Gielen, P. Rombouts, Calibration of DAC mismatch errors in $\Sigma\Delta$ ADCs based on a sine-wave measurement. IEEE Trans. Circuits Syst. II Express Briefs **60**(9), 567–571 (2013)
17. R.G.H. Eschauzier, L.P.T. Kerklaan, J.H. Huijsing, A 100-MHz 100-dB operational amplifier with multipath nested Miller compensation structure. IEEE J. Solid-State Circuits **27**(12), 1709–1717 (1992)
18. T. Wu, C. Jih, J. Chen, C. Wu, A low glitch 10-bit 75-MHz CMOS video D/A converter. IEEE J. Solid-State Circuits **30**(1), 68–72 (1995)

Chapter 6
Two-Step Open-Loop VCO-Based ADC

6.1 Introduction

As seen in Chap. 1, for both wireline and wireless communications, new standards always require larger data rates and signal bandwidths, e.g. VDSL2 and WLAN 802.11n are expected to extend their analog bandwidths to 30 MHz and 40 MHz respectively. In the meantime, more and more work is done with mobile devices powered by battery. Tablets and smart phones with explosive shipments are good examples. Although CMOS technology scaling is beneficial for the increasing bandwidth requirement, more challenges are imposed to the analog and mixed-signal designs, limiting the achievable accuracy and power efficiency. One solution for ADC design is to shift more quantization and A/D conversion from the analog and the voltage domains to the digital and the time domains. In Chap. 4, several new VCO-based quantizer and ADC architectures (both closed-loop and open-loop ones) have been presented. The main design challenge is on how to solve the VCO V-F nonlinearity. In this chapter we explore an alternative new approach.

We combine the two-step conversion principle [1] with a frequency-type VCO-based quantizer [2] to achieve 1st-order noise shaping and solve the VCO nonlinearity issue. The nonlinearities of the coarse and the fine VCO-based quantizers are mitigated by distortion cancellation and input swing reduction schemes. At the same time, the intrinsic DEM of the VCO-based quantizer output relaxes the DAC matching requirement greatly. Because of the simple ADC architecture and the highly digital building blocks, a 42fJ/step ADC figure-of-merit (FoM) is realized.

This chapter covers the full design procedure of this two-step open-loop VCO-based ADC [3, 4]. First the proposed ADC architecture and the system-level design are given. The main ADC nonidealities are considered and modelled to guarantee the ADC performance robustness. Then the circuit design of the frequency-type VCO-based quantizer and the current-steering DAC together with the passive subtractor are presented. Finally the measurement results are shown and discussed.

X. Xing et al., *Design of Power-Efficient Highly Digital Analog-to-Digital Converters for Next-Generation Wireless Communication Systems*, Signals and Communication Technology, https://doi.org/10.1007/978-3-319-66565-8_6

6.2 Architecture Design of Two-Step Open-Loop VCO-Based ADC

In Fig. 4.2 in Chap. 4, the structure and behavioral model of a frequency-type VCO-based quantizer [2] has been shown. First a ring VCO converts the input voltage into phase information, which is illustrated by the rising/falling edges of the VCO outputs. Then a sense-amplifier flip flop (SA FF) is used to synchronize each VCO output, introducing a quantization error in the phase domain. Finally a digital differentiator (constructed by a TSPC FF and an XOR gate) processes the sampled phase and outputs digital frequency information. By applying such a simple, highly digital circuit, not only quantization is done, but also 1st-order noise shaping is obtained. In addition, there is an intrinsic dynamic element matching (DEM) for the digital output. However, the VCO's nonlinear transfer function limits its application in ADC design. Although the 2nd-order harmonics can be cancelled by a pseudo differential topology, the VCO 3rd-order nonlinearity is still a performance limitation for designs with more than 6-bit linearity.

In ADC design, in order to break the exponential increase of the number of comparators occurring in flash ADCs and to reach higher accuracies, the two-step A/D conversion topology has been proposed [1]. The drawback of this two-step architecture is that it finishes one conversion in two clock periods, hence introducing a delay of one clock period. Here a DAC and an analog subtractor with full accuracy are required to generate the residual signal. Also a full accuracy analog amplifier is needed to amplifier the residual signal.

In this section, we combine the two-step conversion principle with a VCO-based quantizer to achieve 1st-order noise shaping and to solve the VCO nonlinearity issue. At the same time, the intrinsic DEM of the VCO-based quantizer output relaxes the DAC matching requirement greatly.

6.2.1 A Two-Step Open-Loop VCO-Based ADC Architecture

6.2.1.1 Intrinsic VCO Nonlinearity Mitigation

The proposed two-step open-loop VCO-based ADC architecture and its illustrative signal spectra are shown in Fig. 6.1 (VCO-based quantizers have been presented in Fig. 4.2 in Chap. 4). Its structure and operation principle are similar to a two-step flash ADC except for two differences. First, highly digital frequency-type VCO-based quantizers with 1st-order noise shaping and intrinsic DEM are used instead of simple flash quantizers. Secondly, there is no analog amplifier preceding the fine quantizer.

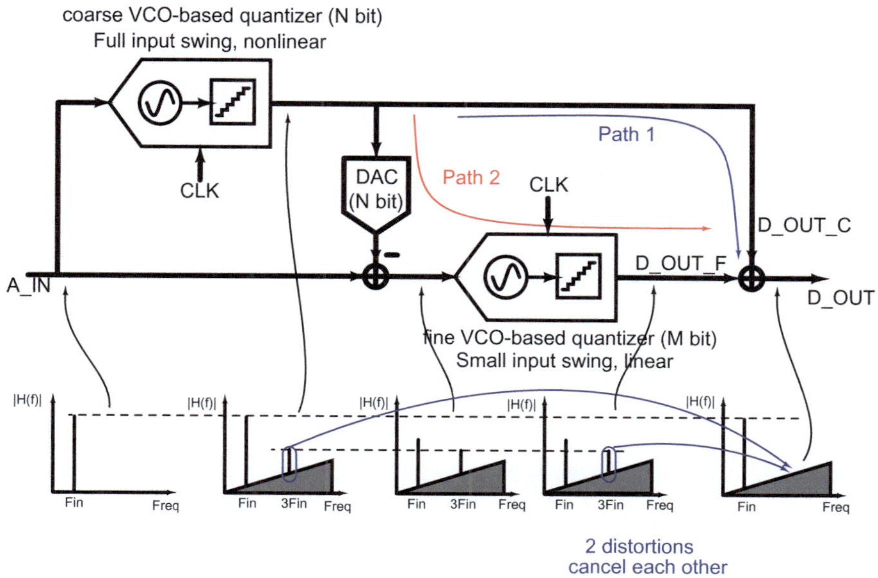

Fig. 6.1 Proposed two-step open-loop VCO-based ADC architecture and the VCO nonlinearity mitigation concept

In the two-step open-loop ADC, the N-bit VCO-based coarse quantizer converts the analog input into digital bits. With full-swing input voltage, the coarse quantizer's digital output includes not only the desired signal, but also 1st-order shaped quantization error and harmonic distortions. A N-bit DAC is used to convert the coarse quantizer output back to an analog signal, without introducing quantization noise. By subtracting the DAC output from the ADC input, the residual error of the coarse quantizer is obtained. Since the subtractor is linear, the noise and distortion parts of the residual signal are almost the same as those of the coarse output. If the number of coarse quantization bits is large enough and so the coarse quantization step is small enough, the voltage swing of the residual signal is much smaller than the whole ADC's input signal swing. The analog residual signal is converted further by the M-bit fine VCO-based quantizer. With its low input signal swing, the fine VCO-based quantizer transfer function is very linear and its nonlinearity does not limit the ADC performance. In the fine quantizer, although the input signal swing is small, because of the integration provided by the VCO and the quantization in the time domain, all $2^M - 1$ quantization steps are used [5] and only a small quantization error is introduced. Finally, the digital outputs of the coarse and the fine VCO-based quantizers are summed as the output of the whole ADC.

In this design, the gain of the path 2 in Fig. 6.1 (the DAC, the subtractor and the fine VCO-based quantizer) is designed to be 1 to match that of the path 1. By doing this, all nonidealities (quantization error, thermal noise, distortion) of the coarse VCO-based quantizer are cancelled, and the two-step open-loop VCO-based ADC is equivalent to a linear 1st-order Delta-Sigma modulator (DSM) with M-bit quantization. The noise contributions of the DAC and the VCO in the fine quantizer refer to the ADC input directly, and they need to be optimized in the circuit-level design.

6.2.1.2 Implicit DAC DEM

For the DAC in Fig. 6.1, its input swing is almost full and there is no gain preceding it, so its mismatch refers to the ADC input directly. If a simple flash ADC is used as the coarse quantizer, there is no implicit DEM for the D/A conversion. As a result, either transistor intrinsic matching or an explicit DEM block is required to achieve the required linearity. Intrinsic transistor matching means a larger area and a complex switching scheme, consuming large chip area and power. With an external DEM, not only extra power consumption and chip area are required, but also extra path delay is introduced, which affects the two-step ADC performance and sets more stringent requirements on the other blocks.

In our two-step open-loop ADC, a frequency-type VCO-based quantizer is applied as coarse quantizer instead of a flash ADC. In such a quantizer, the different phase outputs of the ring VCO change their state consecutively, and a state change corresponds to a "1" in the frequency detector output. This means that there is an intrinsic DEM with the quantizer digital outputs [2]. For the DAC design, because of the intrinsic DEM of the digital input, the DAC cell matching requirement is relaxed and can easily be reached. With this arrangement, the DAC nonlinearity is no longer the ADC linearity bottleneck. Furthermore, there is no extra hardware or power consumed, and no extra delay is introduced in the path.

6.2.1.3 Implicit AAF

Another benefit of using a VCO-based quantizer as the coarse and fine quantizers is that intrinsic 1st-order anti-aliasing filtering (AAF) is kept in the proposed two-step open-loop VCO-based ADC. It has been explained in Chap. 4 that there is intrinsic 1st-order AAF with the frequency-type VCO-based quantizer [6]. From the analog input of the open-loop ADC to its digital output, there are in total three paths, and in each path there is a VCO-based quantizer preceding another block (DAC, digital adder). With the VCO-based quantizer as the front-end, the high-frequency signal component is attenuated by 1st-order filtering in every path, so this AAF is done for the whole ADC.

To verify the proposed two-step open-loop VCO-based ADC architecture, a 40 MHz-BW 10-bit ADC has been designed and implemented. In this design, both the coarse and the fine VCO-based quantizers are 5-bit pseudo-differential structures (composed by 2 4-bit single-end ones). The sampling frequency of the two-step open-loop VCO-based ADC is 1.6 GHz, with an oversampling ratio (OSR) of 20.

6.2.2 *Nonidealities of the Two-Step Open-Loop VCO-Based ADC*

In a real circuit implementation, there are several nonidealities in each block, such as gain variation, delay and mismatch. In reality, the ADC clock signal can not be jitter-free either. To obtain a stable ADC performance and to determine the design specifications for each building block, important ADC nonidealities should be considered and included in the behavioral-level simulations.

As the most important error source and design objective in VCO-based ADC design, the VCO V-F nonlinearity is modelled here by a 4th-order polynomial:

$$f = f_o + k_{vco}A_{in} + k_2A_{in}^2 + k_3A_{in}^3 + k_4A_{in}^4 \tag{6.1}$$

where the coefficients are calculated reversely from the distortion levels in SPICE simulation results:

$$HD_2 = \frac{1}{2}\frac{k_2}{k_{vco}}A_{in} \quad HD_3 = \frac{1}{4}\frac{k_3}{k_{vco}}A_{in}^2 \quad HD_4 = \frac{1}{8}\frac{k_4}{k_{vco}}A_{in}^3 \tag{6.2}$$

6.2.2.1 Block Gain Variation

In the frequency-type VCO-based quantizer, the quantizer gain is determined by the gain between the input control voltage and the oscillation frequency. Once the delay cells are fixed, such gain is decided by the transconductance of the current source transistor. The gain of a current-steering DAC is set by its bias current. In a CMOS implementation, both the VCO-based quantizer gain and the DAC gain can vary as much as $[-20\%, 20\%]$. The gain deviation of the coarse and the fine VCO-based quantizers and the DAC have different mechanisms and effects on the ADC performance.

From the ADC analog input to the fine quantizer input, there are two paths: one analog path with a gain of unity, and another one constructed by the coarse quantizer and the DAC. If the gain of the second path is not exact unity, then the voltage swing of the residual signal will become larger. With a larger input signal swing, the linearity of the fine quantizer will be worse. The nonlinearity of the fine quantizer refers to the ADC input directly. From the coarse quantizer output to the whole ADC output, there are also two paths: one digital path with unity gain (path 1 in Fig. 6.1), and the path composed by the DAC and the fine quantizer (path 2 in Fig. 6.1). If the gains of paths 1 and 2 are not the same, then part of the coarse quantizer output appears in the ADC output. Although the coarse quantization is also 1st-order noise shaped, its input swing is almost full swing and its distortion leakage degrades the ADC linearity performance.

In Fig. 6.2, the sensitivity of the ADC SNDR performance to the block gain variations are shown. The ADC performance is robust to the coarse ADC gain variation:

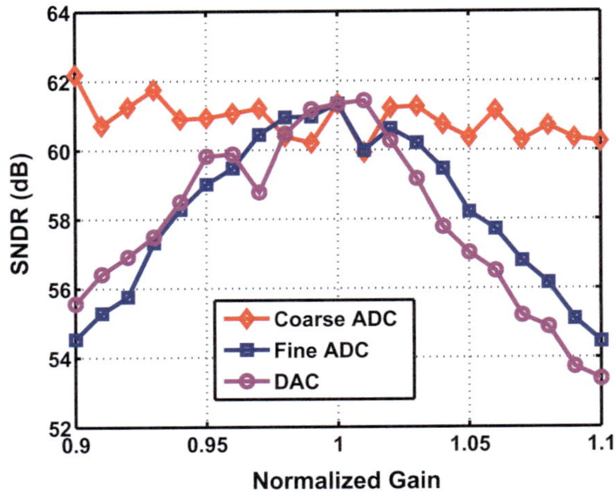

Fig. 6.2 ADC SNDR performance as a function of the normalized gain of the coarse ADC, fine ADC and DAC

within [−10%, 10%] gain variation, the gain variation has almost no effect on the ADC performance. However, the ADC SNDR is much more sensitive to the gain variations of the DAC and the fine ADC: smaller than ±5% gain variation windows are required for the ADC to achieve 60 dB SNDR. Such simulation results show that the second mechanism is much more severe than the first one. In circuit design, the transconductances of current source transistors in the coarse and the fine VCOs are digitally adjustable, and so are their gains. Also the DAC bias current can be tuned continuously. With these discrete and continuous tunings, the gain of path 2 can be tuned to be very close to 1.

6.2.2.2 Block Delay

In the frequency-type VCO-based quantizer, the input-to-output and the clock-to-output delays of all digital gates contribute to the total quantizer digital delay. For the NRZ current-steering DAC, the finite settling of the output current introduces time delay.

For the two-step open-loop VCO-based ADC, its performance not only depends on the block gains, but also on the delay of each block. Since even building block delay causes time/phase mismatch between two paths, increasing the residual signal swing and degrading the distortion cancellation. In Fig. 6.3, the relationships between the ADC SNDR and the delays of the coarse ADC and the DAC are shown. For these two blocks, less than $0.25{\cdot}T_s$ delay is allowed for the ADC to reach 60 dB SNDR. On the other hand, because a digital signal is robust to time skew, the fine ADC can tolerate as much as one clock period delay. The reason why the coarse ADC delay tolerance is smaller is that its digital output needs to be processed further by another block to generate the analog residual signal, and the analog signal is much more sensitive to time delay.

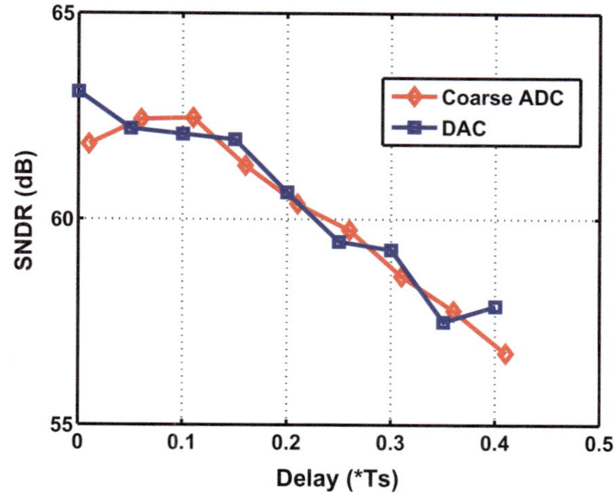

Fig. 6.3 ADC SNDR performance as a function of the delays of the coarse ADC and the DAC

6.2.2.3 DAC Mismatch

In an N-bit DAC design, because of the systematic and random process variations, the $(2^N - 1)$ DAC cell values can not be exactly the same. This mismatch makes the DAC transfer function nonlinear. In our design, the DAC nonlinearity is not suppressed by any gain and appears in the ADC output directly. Fortunately, by using a frequency-type VCO-based quantizer as the coarse quantizer, the DAC input is intrinsic DEMed, which greatly relaxes the matching requirement on the DAC cells. In Fig. 6.4, the ADC SNDR is plotted as a function of the DAC cell variation standard deviation. For comparison, the ADC SNDR values are also given for the case of no DEM. In our design with DEM, the DAC cell sigma can be as large as 0.15 unit, which can be realized easily in CMOS technology. If there is no DEM, then the ADC linearity performance is only around 37 dB, and 0.005 standard deviation is required to reach 60 dB performance. It is obvious that the proposed application of the coarse VCO-based quantizer largely relaxes the DAC design.

6.2.2.4 Clock Jitter

Another important nonideality for ADC design is clock jitter. In the proposed two-step open-loop VCO-based ADC, clock jitter introduces an error in all three blocks. In the VCO-based quantizer, because the VCO integration is preceding the time-domain sampling, the clock-jitter-induced noise is 1st-order shaped. On the other hand, the error caused by the DAC clock jitter is referred to the ADC input directly. In Fig. 6.5, the ADC performance degradation due to clock jitter is presented. Because of the NRZ DAC topology and the 5-bit coarse quantization, as high as 15ps clock jitter can be tolerated by our design to reach 60 dB SNDR.

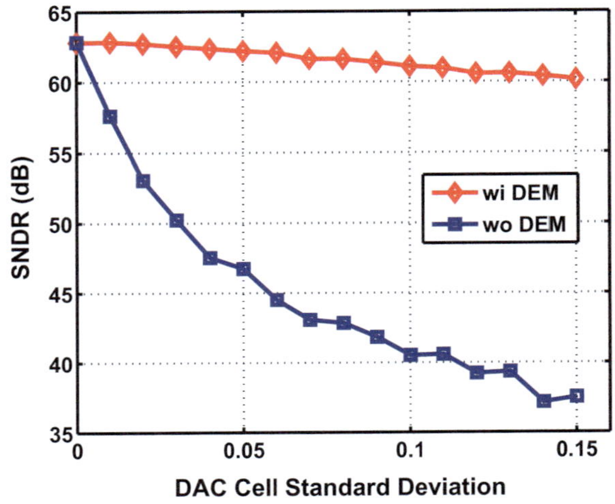

Fig. 6.4 ADC SNDR performance as a function of the standard deviation of the current of the DAC cells. For comparison, both cases with and without DEM are showed

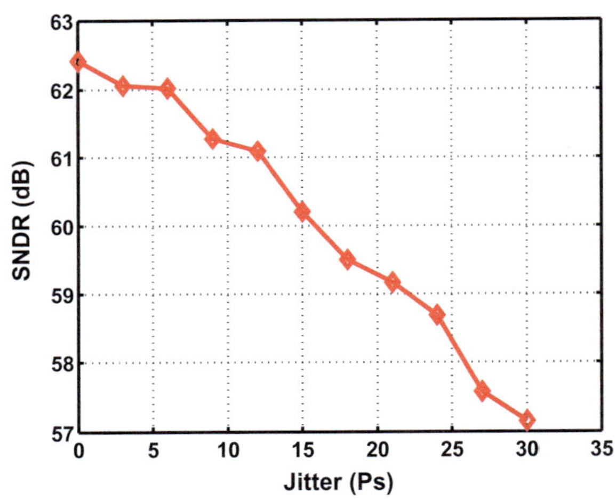

Fig. 6.5 ADC SNDR performance as a function of the clock jitter: around 15ps jitter can be tolerated for 10-bit accuracy

6.3 Circuit Implementation of the Two-Step Open-Loop VCO-Based ADC

To design the two-step open-loop VCO-based ADC, two VCO-based quantizers, one DAC and one subtractor are required. Theoretically speaking, since the nonidealities of the coarse VCO-based quantizer are cancelled, its design (for example phase noise) can be relaxed. However, in frequency-type VCO-based quantizers, most of power consumption is from the digital gates and not from the VCO. So it is not so beneficial to design two different VCO-based quantizers. Based on this fact and to reduce the design time, the coarse and fine VCO-based quantizers are taken the same in this design.

6.3.1 VCO-Based Quantizer Design

To cancel the even-order harmonic distortions, the pseudo-differential 5-bit VCO-based quantizers are composed of 2 4-bit single-ended quantizers. The schematic of the 15-stage ring VCO of a 4-bit single-end quantizer is presented in Fig. 6.6, which is composed of a ring of delay cells and a transconductor. The circuit of the delay cell is shown in the blue dashed box [7]. It is a pair of current-starved inverters, together with two more inverters to synchronize their outputs. The oscillation frequency of the ring VCO is determined by the drain currents of the delay cells, which is generated by the voltage-controlled current source. To accommodate the corner variations introduced by circuit fabrication, a 3-bit binary control scheme is adopted to tune the transconductance of the current source, and hence the VCO gain.

With this topology, the VCO output can not be full swing (it can reach the power supply but it can not reach the ground), especially when the oscillation frequency is high. To detect the edge of the VCO phase output correctly, an SA FF with NMOS input is used as the time-domain S/H. With a 500 mV common-mode voltage and a 400 mVpp input swing, the maximum oscillation frequency of the ring VCO is designed to be no larger than half of the sampling frequency (800 MHz) to avoid overloading the quantizer in the time domain.

The simulated power supply sensitivity of the VCO-based quantizer gain is 13.1%/V, and its temperature sensitivity is 2.1 ‰ (per-mille). In Fig. 6.7, the transistor-level simulations of the VCO-based quantizer with 0dBFS input power are shown. Compared to the 4-bit single-ended topology (left), the 5-bit pseudo-differential topology not only increases the quantization accuracy, but it also cancels the even-order harmonic distortions. As presented, the SNDR of the 5-bit pseudo differential VCO-based quantizer is about 37 dB. In Fig. 6.8, the simulation result of the VCO-

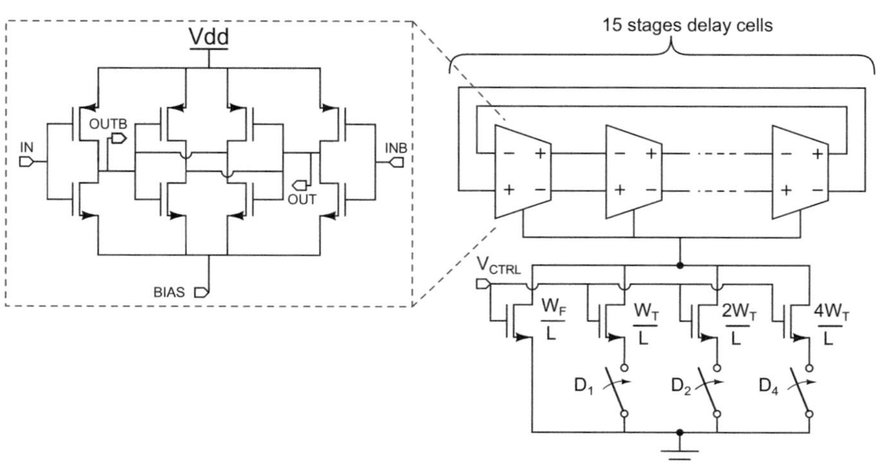

Fig. 6.6 Schematic of the 15-stage ring VCO in the quantizer (single-ended); the delay cell circuit is shown in the *blue dashed box*

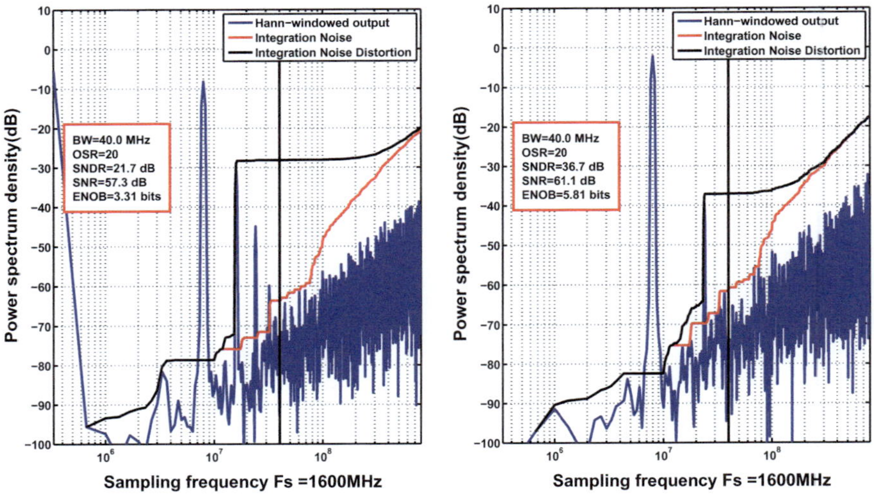

Fig. 6.7 Large-signal simulation of a 5-bit frequency-type VCO-based quantizer. Both the single-ended result (*left*) and the pseudo-differential result (*right*) are presented

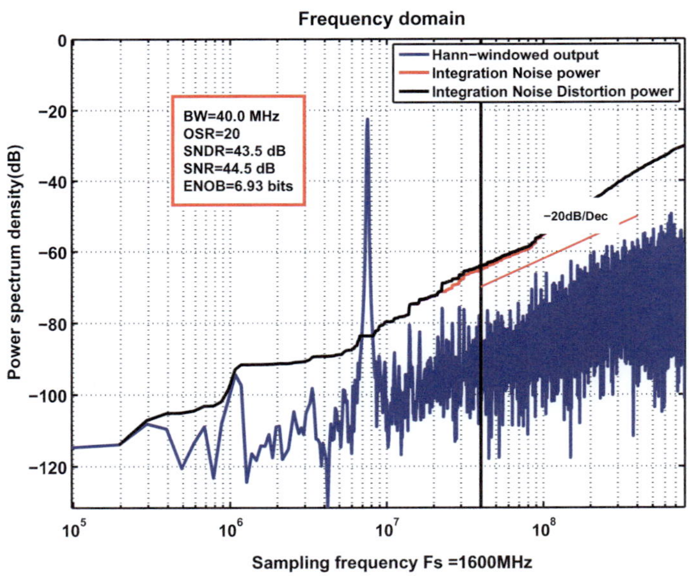

Fig. 6.8 Small-signal simulation of a 5-bit frequency-type VCO-based quantizer; with around −20 dBFS input, the 3rd-order distortion is about −76 dB

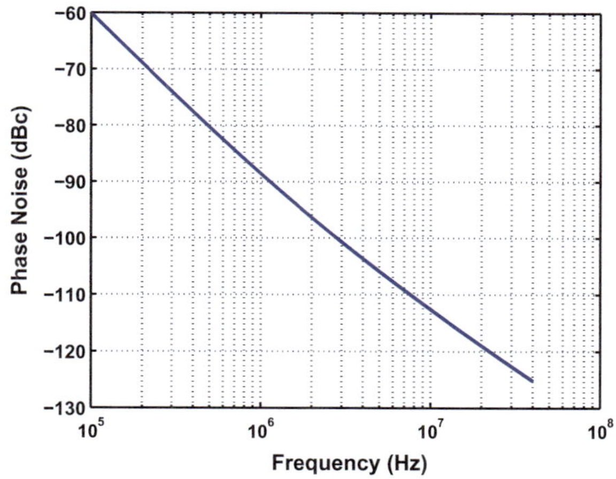

Fig. 6.9 The simulated VCO phase noise; it is −113 dBc/Hz at an offset frequency of 10 MHz

based quantizer with small-signal input is presented. When the input is reduced to −22 dBFS, the VCO-based quantizer is quite linear, and the 3rd-order distortion is only around −76 dBFS (Fig. 6.8).

The simulated phase noise of the VCO is shown in Fig. 6.9, which is about −113 dBc/Hz at 10 MHz offset frequency. According to the formula in [8], the calculated signal-to-phase-noise-ratio is about 68 dB. It is clear that this VCO phase noise does not limit the ADC targeted performance (10bit).

6.3.2 DAC and Subtractor Design

The 5-bit DAC used in the two-step open-loop VCO-based ADC is a NRZ current-steering DAC, chosen for its high speed and robustness to clock jitter. The detailed schematic of the DAC together with the subtractor is shown in Fig. 6.10. The basic DAC cell is composed of a current-source transistor, a cascode transistor and a pair of switches with full-swing digital input. For the current-source transistor, about 0.64 μm^2 active area is used to achieve the required matching (around 8% standard deviation). For the cascode transistor, 2× minimum gate length is used to reach about 7.5× gain. This gain not only provides sufficient shielding from the noisy switching node to the quiet bias node, but also increases the DAC output impedance. Because of the full-swing input, the size of the switch transistors can be minimized, reducing the coupling effect. With proper transistor sizes and source terminal voltages, even with a full-swing DAC input, the switch transistor is still in the saturation region, increasing the DAC output impedance further. To keep the CM voltages for the coarse and the fine VCO-based quantizers the same, two CM current sources are added to the DAC, which are PMOS cascode constant current sources. The subtraction operation

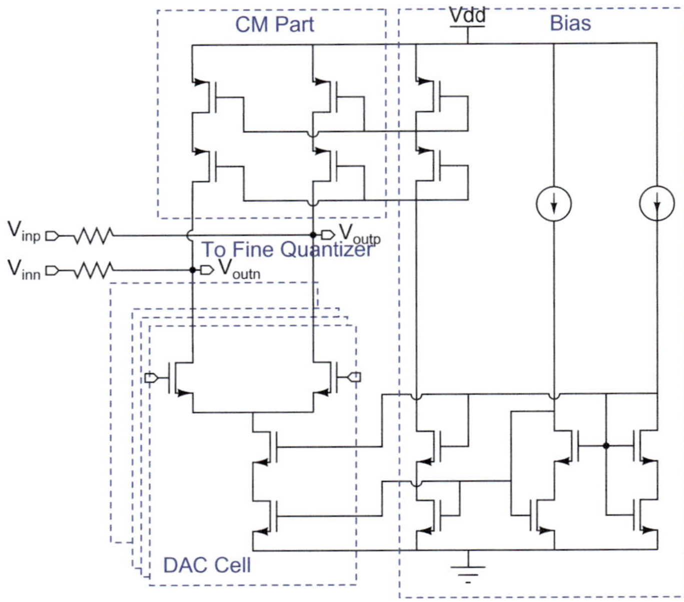

Fig. 6.10 Schematic of the current-steering DAC; the resistor connected to the ADC input and the DAC output forms a passive subtractor

Fig. 6.11 Noise simulation of the feedback current-steering DAC and the passive subtractor

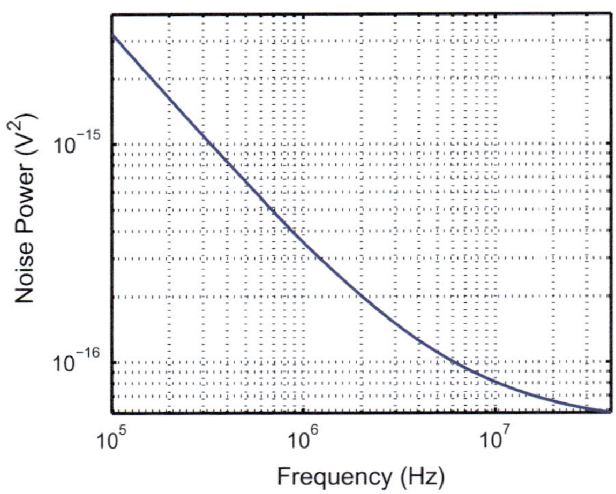

between the positive and the negative parts of the coarse VCO-based quantizer are realized by changing the polarity of the DAC's negative part output.

To avoid active power consumption, a simple passive subtractor is applied in this design, by connecting resistors between the ADC inputs and the DAC outputs. The output currents of the current-steering DAC go through the resistors and cause corresponding voltage drops across the resistors. The ADC differential input voltage

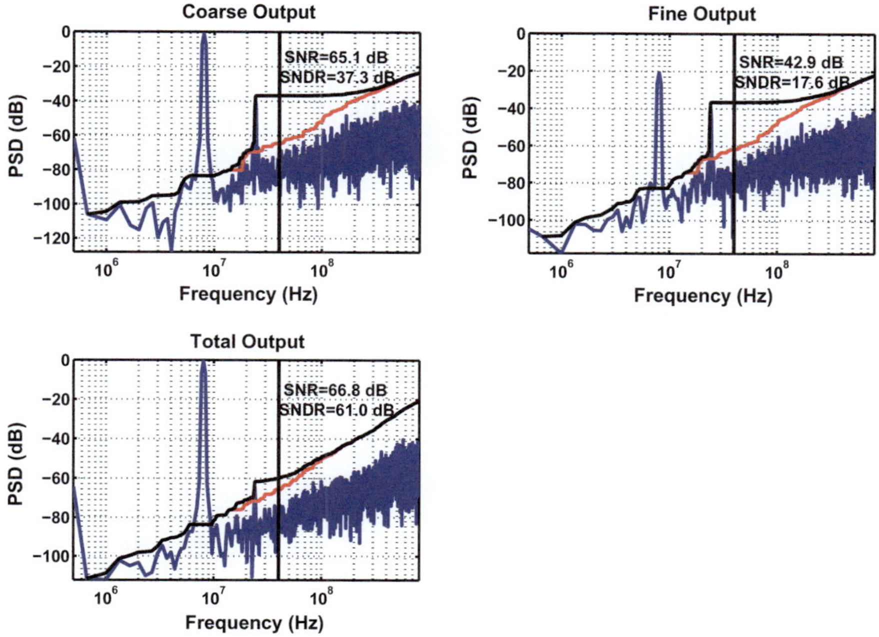

Fig. 6.12 Transistor-level simulation result of the two-step open-loop VCO-based ADC. *Top left* PSD plot of the coarse quantizer output; *top right* PSD plot of the fine quantizer output; *bottom left* PSD plot of the overall ADC output

minus the voltage drop results in the residual signal for the fine VCO-based quantizer. As the load of the current-steering DAC, the resistor value is 500 Ω, which means that the DAC full-scale current is 0.4 mA, and the DAC cell current is around 26.7 μA. Noise simulation results are presented in Fig. 6.11, showing that the RMS (root-mean-square) output noise voltage of the current-steering DAC and the subtractor is about 54 V, which corresponds to about −77 dBFS.

With the above designed building blocks, the two-step open-loop VCO-based ADC is constructed and simulated in Cadence APS. The simulation results are shown in Fig. 6.12. In the top left, the output PSD of the coarse quantizer is shown: 65.1 dB SNR is achieved, but the SNDR is only 37.3 dB. The PSD of the fine quantizer output is given in the top right: compared to the coarse quantizer output, the signal energy is 20 dB lower and the 3rd-order distortion energy is the same. In the whole ADC output (PSD shown in bottom left), the distortions of the coarse and the fine quantizer outputs cancel each other and 61dB SNDR is obtained.

6.4 Experimental Results and Discussions

The proposed two-step open-loop VCO-based ADC has been fabricated in TSMC 40 nm general purpose (GP) CMOS. The die photo is given in Fig. 6.13; the building blocks are marked by red rectangles. Because of the simple and highly digital architecture, only 0.017 mm^2 core area is utilized. To clearly illustrate the two-step conversion idea, both the digital outputs of the coarse and the fine VCO-based quantizers are converted into binary codes and fed to digital buffers and a low-voltage differential signal (LVDS) interface. The remaining space is occupied by on-chip decoupling capacitors (for power supplies and bias voltages) and bonding pads.

To measure the two-step open-loop VCO-based ADC, analog sine voltages with 12 MHz frequency, 500 mV offset voltage and variable amplitude are used as the ADC input. The measured results with −1 dBFS input power are shown in Fig. 6.14. In the top left of Fig. 6.14, the power spectral density (PSD) of the coarse quantizer output is plotted, the 20 dB/decade roll-off characteristic provided by the VCO inherent integration is evident. With almost full input swing and the nonlinear VCO transfer function, the 3rd-order distortion appears, and the linearity of the quantizer is about 5-bit. The actual matching between the positive and the negative parts of the pseudo differential quantizer is limited in practice, so part of the 2nd-order distortion still remains. In the top right of Fig. 6.14, the PSD of the fine quantizer output is shown. Here the signal power is about 20dB lower than that of the coarse quantizer output, and the distortions are accurately the same as is the coarse quantizer output. As presented in the simulation result of Fig. 6.8, with such a small input power, the transfer function of the fine VCO-based quantizer is very linear. So the distortion of the fine quantizer output is from the residual signal itself and not from the fine

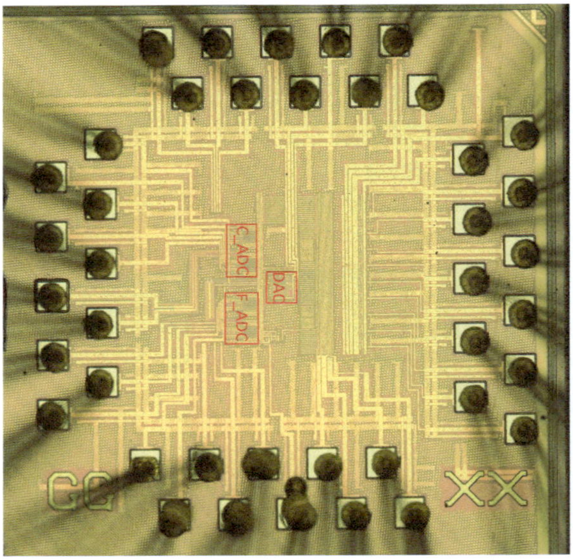

Fig. 6.13 Die photo of the two-step open-loop VCO-based ADC

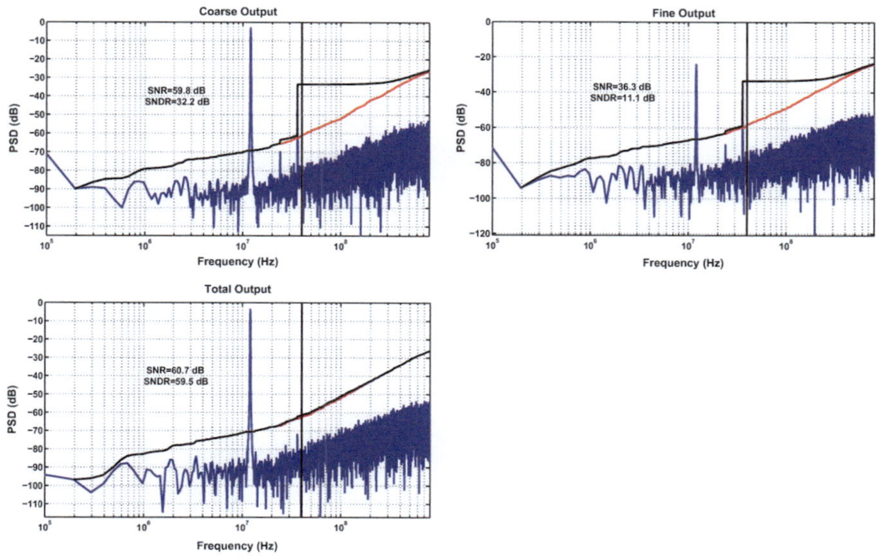

Fig. 6.14 Measured results of the two-step 1st-order VCO-based DSM with −1 dBFS 12 MHz sine input. *Top left* PSD plot of the coarse quantizer output; *top right* PSD plot of the fine quantizer output; *bottom left* PSD plot of the overall ADC output

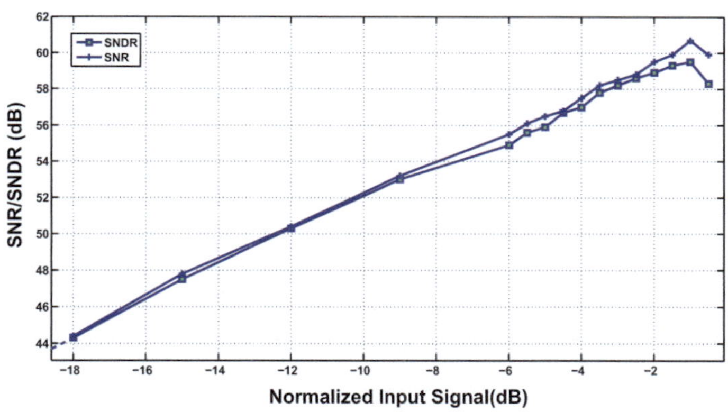

Fig. 6.15 The measured ADC SNR/SNDR as a function of the analog input power

quantizer. Finally, by adding the coarse and fine quantizer outputs together, their distortions cancel each other. As a result, 59.5 dB SNDR and 67.6 dB spurious-free dynamic range (SFDR) are achieved for the whole ADC, as shown in the bottom left of Fig. 6.14. The relationship between the measured ADC SNR/SNDR and the analog input power is given in Fig. 6.15.

The power consumption of the two-step VCO-based ADC is about 2.57 mW, most of which is from the digital gates (FF, XOR). Because of the simple open-loop architecture and the highly digital building blocks, an excellent FoM of 42fJ/Step

Table 6.1 Measurement summary of the two-step open-loop VCO-based ADC

Specification	Value
Signal bandwidth	40 MHz
Sampling frequency	1.6 GHz
Peak SNR	60.7 dB
Peak SNDR	59.5 dB
Power	2.57 mW
FoM_1	42fJ/Step
FoM_2	161.4 dB
Core area	0.017 mm^2
Technology	TSMC 40 nm GP

((Power/(2BW$*2^{ENOB}$))) or 161.4 dB (SNDR+10log(BW/Power)) is achieved, indicating the high energy efficiency of this structure. Since the sampling frequency and the power consumption are limited by digital blocks, it is expected that with a more advanced CMOS technology, a higher ADC bandwidth (or higher ADC accuracy) and an even lower FoM can be achieved. The measurement results of the two-step open-loop VCO-based ADC are summarized in Table 6.1, and will be compared to the state-of-the-art works at the end of Chap. 8, after having presented the final ADC design.

6.5 Conclusions

A frequency-type VCO-based quantizer has several advantages and is very suitable for ADC design, offering high power efficiency, intrinsic 1st-order noise shaping and implicit DEM. To utilize these merits and to mitigate the VCO nonlinearity, a two-step open-loop VCO-based ADC architecture has been proposed and implemented in this chapter. With compact open-loop structure and highly-digital building blocks, a stable performance, a high bandwidth, a very low FoM and a small area can easily be achieved. The nonlinearities of the coarse and the fine VCO-based quantizers are solved by distortion cancellation and input swing reduction schemes respectively. The DAC matching is relaxed greatly as its input has intrinsic DEM.

In this chapter the effects of the different nonidealities have been investigated in detail, based on which the design specifications for each building block have been determined. Then the full circuit design and the implementation of the two-step open-loop VCO-based ADC in 40 nm CMOS has been presented. Simulation results of all building blocks and the whole ADC have been given in detail. Measurement results show that the proposed ADC achieves a SNDR of 59.5 dB and a SFDR of 67.6 dB for a 40 MHz bandwidth, with a 1.6 GHz sampling frequency and only 2.57 mW power consumption. The corresponding excellent FoM (42fJ/Step) is one of the best among high-bandwidth ($\geq$20 MHz) DSM ADCs. This VCO-based ADC design has achieved a high-digital architecture and thus a outstanding power efficiency, however, the first-order noise-shaping limits its accuracy performance. Therefore, in the next chapter, we will present a second-order VCO-based ADC design, targetting better accuracy and even better power efficiency.

References

1. H. van der Ploeg, R. Remmers, A 3.3-V, 10-b, 25-MSample/s two-step ADC in 0.35-μm CMOS. Solid-State Circuits, IEEE J. **34**(12), 1803–1811 (1989)
2. M.Z. Straayer, M.H. Perrott, A 12-bits, 10-MHz bandwidth, continuous-time Delta-Sigma ADC with a 5-bits, 950-MS/s VCO-based quantizer. IEEE J. Solid-State Circuits **43**(4), 805–814 (2008)
3. X. Xing, P. Gao, G. Gielen, A 40 MHz-BW two-step open-loop VCO-based ADC with 42fJ/step FoM in 40 nm CMOS, in *Proceedings of IEEE ESSCIRC*, (Bucharest, Romania, 2013), pp. 327–330
4. X. Xing, G.G.E. Gielen, A 42fJ/Step-FoM two-step VCO-based Delta-Sigma ADC in 40 nm CMOS. Solid-State Circuits, IEEE J. **50**(3), 714–723 (2015)
5. M. Park, M.H. Perrott, A 78 dB SNDR 87 mW 20 MHz bandwidth continuous-time $\Delta\Sigma$ ADC with VCO-based integrator and quantizer implemented in 0.13 μm CMOS. Solid-State Circuits, IEEE J. **44**(12), 3344–3358 (2009)
6. K. Reddy, S. Rao, R. Inti, B. Young, A. Elshazly, M. Talegaonkar, P.K. Hanumolu, A 16-mW 78-dB SNDR 10-MHz BW CT $\Delta\Sigma$ ADC using residue-cancelling VCO-based quantizer. Solid-State Circuits, IEEE J. **47**(12), 2916–2927 (2012)
7. G. Taylor, I. Galton, A mostly-digital variable-rate continuous-time Delta-Sigma modulator ADC. Solid-State Circuits, IEEE J. **45**(12), 2634–2646 (2010)
8. J. Kim, T.K. Jang, Y.G. Yoon, S.H. Cho, Analysis and design of voltage-controlled oscillator based analog-to-digital converter. IEEE Trans. Circuits Syst. I Regul. Pap. **57**(1), 18–30 (2010)

Chapter 7
VCO-Based 0-$\Delta\Sigma$ MASH ADC

7.1 Introduction

As discussed in Chap. 4, to address the issue of limited voltage headroom and to take advantage of the reduced CMOS gate delay in scaled CMOS technologies, VCO-based highly-digital ADCs have recently emerged in recent years. However, the performance of VCO-based ADCs is limited by the inherent nonlinearity of the VCOs. As shown in Fig. 7.1, a VCO-based ADC can be utilized as a quantizer in a single-loop $\Delta\Sigma$ modulator. The preceding loop filter suppresses both the quantization noise and the nonlinearity, however, a high loop gain results in a high power consumption and potential instability problems. As discussed in Chap. 3, in traditional $\Delta\Sigma$ ADC design, compared to single-loop structures, MASH architectures can offer a higher-order noise shaping with less instability concerns. This structure can also be implemented with VCO-based quantizers as will be demonstrated.

In this chapter, the VCO-based MASH $\Delta\Sigma$ ADC architecture will be investigated. Specially, an open-loop quantizer without preceding filters, or so-called a zero-order quantizer, can be adopted in the stages of the MASH modulator. With VCO-based quantizers, the 0-$\Delta\Sigma$ MASH topology can overcome the VCO nonlinearity problem due to its inherent operation principle, being more power efficient than the single-loop structure. Section 7.2 will analysis the architecture and the nonlinearity-cancellation technique, with a 0–2 MASH VCO-based ADC as an example [1, 2]. The system-level design and the nonlinearity-cancellation robustness against gain variations will be investigated. Furthermore, the delay matching technique for the nonlinearity cancellation will be presented. Section 7.3 will cover the circuit design of the 0–2 MASH VCO-based ADC building blocks, including the three-input adder, the VCO-based quantizer, the analog integrator, the feedback DAC and the digital interface. The measurement results in 40-nm CMOS and discussions will be given in Sect. 7.4. At last, the conclusions will be provided in Sect. 7.5.

X. Xing et al., *Design of Power-Efficient Highly Digital Analog-to-Digital Converters for Next-Generation Wireless Communication Systems*, Signals and Communication Technology, https://doi.org/10.1007/978-3-319-66565-8_7

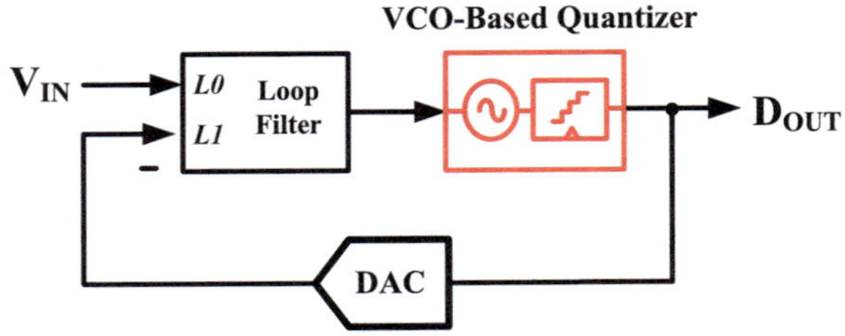

Fig. 7.1 VCO-based quantizer in a single-loop $\Delta\Sigma$ modulator

7.2 Architecture Analysis of the 0-$\Delta\Sigma$ MASH VCO-Based ADC

7.2.1 0-$\Delta\Sigma$ MASH VCO-Based ADC

A 0-order stage is an open-loop quantizer without preceding filters. By replacing flash quantizers with VCO-based quantizers, a 0-$\Delta\Sigma$ VCO-based MASH structure is shown in Fig. 7.2. Several advantages are obtained by adopting this topology.

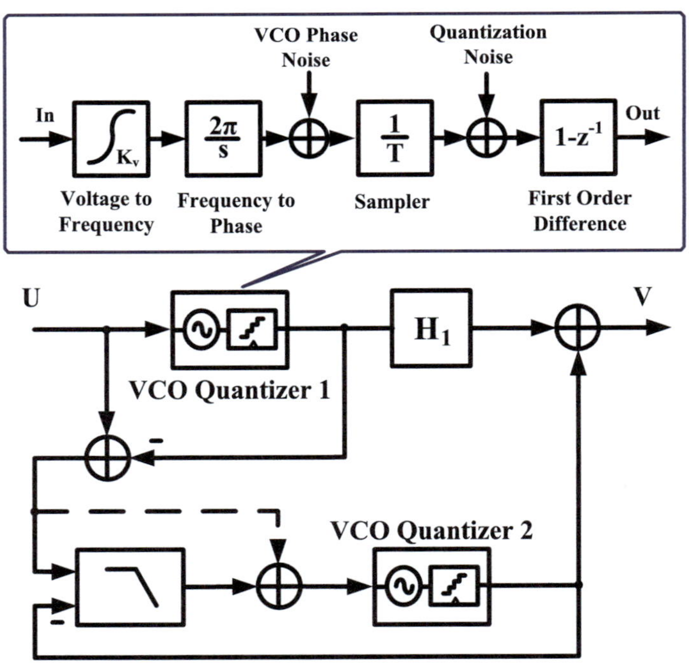

Fig. 7.2 Block diagram of a 0-$\Delta\Sigma$ MASH VCO-based ADC

Firstly, the VCO-based quantizer has an intrinsic 1st-order noise shaping, as evidenced by the 1st-order difference function shown in the behavioral model. Different from other 0-$\Delta\Sigma$ MASH ADCs, e.g., [3–5], which use a flash or a SAR ADC as the 1st-stage quantizer, the first stage of the VCO-based MASH ADC is no longer of zero order. Therefore, any noise leakage due to unperfect matching is 1st-order shaped. As the 1st-order first stage can not provide additional noise shaping for the whole ADC, we still call it 0-$\Delta\Sigma$ MASH structure for simplicity.

Secondly, due to its mostly-digital structure, the VCO-based quantizer is extremely power-efficient. Although the final NTF order is only determined by the second stage, the power consumption of the first stage is reduced.

Thirdly, the VCO-based quantizer has an implicit dynamic element matching (DEM) which relaxes the matching requirements of the DAC cells, and an intrinsic 1st-order anti-aliasing filter, which is lost in the traditional zero-order first stage.

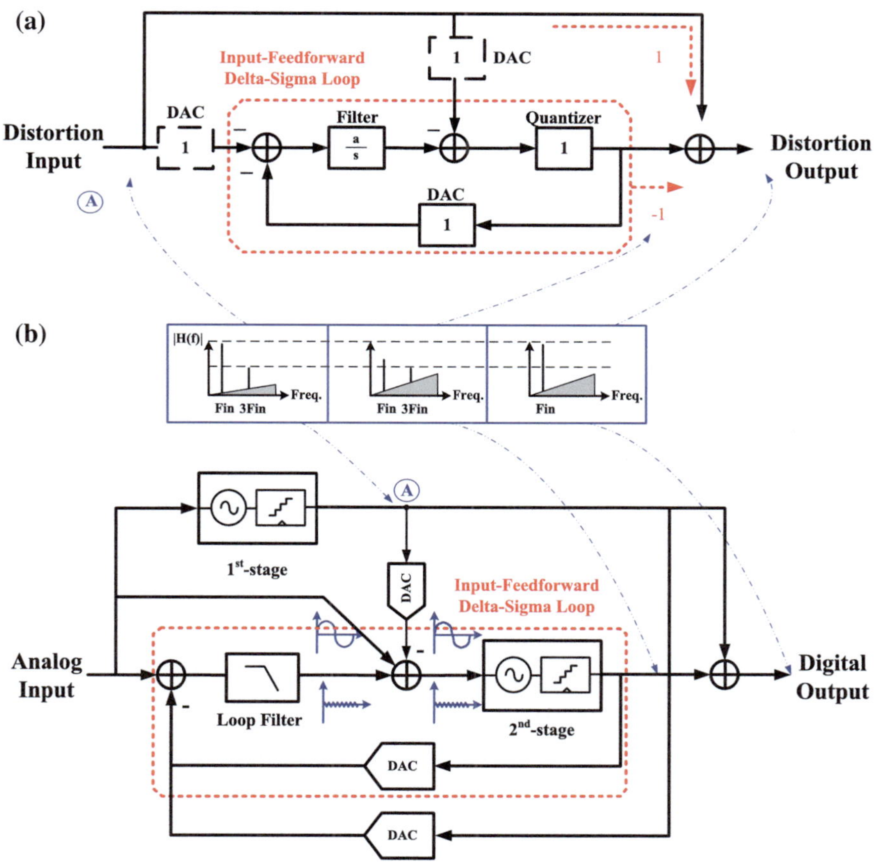

Fig. 7.3 The nonlinearity-cancellation principle in a 0-$\Delta\Sigma$ MASH VCO-based ADC

Most importantly, the VCO nonlinearity problem can easily be solved in this structure. Similar to the quantization noise, the nonlinearity of the first stage can also be cancelled in the cascaded structure [6].

Figure 7.3 illustrates the nonlinearity-cancellation principle, where node A in Fig. 7.3b corresponds to the distortion input position in Fig. 7.3a. For modeling simplicity, H(s) represents the transfer function for the loop filter, while the quantizer and the DACs are represented by blocks with unity gain.

In order to cancel out the distortion, a negative path with gain equal to -1 needs to be created. Particularly, an input-feedforward topology in the second stage can be used to ensure this requirement. The negative path is a conventional input-feedforward $\Delta\Sigma$ loop of which the STF is equal to 1. When the outputs from the positive and the negative paths are summed together, the input distortion is cancelled out thoroughly.

In addition, as the second stage only processes the noise-shaped quantization noise, the input-voltage swing is greatly reduced, resulting in a linear second-stage quantizer [7, 8]. Therefore, the overall output can be expressed as:

$$V = H_1 \cdot U + (H_1 - STF_2) \cdot [(1 - z^{-1})E_1 + D_1] - NTF_2 \cdot E_2 \tag{7.1}$$

where H refers to the stage transfer function, E indicates the quantization noise and D represents all the nonlinear harmonics. Clearly, if the digital filter $H_1 = STF_2$ is adopted, then both the quantization noise and the nonlinearity of the first-stage modulator can be eliminated. As discussed above, the simplest choice to satisfy the condition of $H_1 = STF_2$ is to set $STF_2 = 1$.

Figure 7.4 illustrates the signal model of the 0-$\Delta\Sigma$ MASH VCO-based ADC, where E_1 and E_2 represent the corresponding quantization noise of the two quantizers, K_{Q1} and K_{Q2} indicate the quantizer gains, D_{Q1} represents all the nonlinear harmonics generated in the first quantizer due to VCO's nonlinear behavior, H(s) represents the transfer function of the analog integrator, $K_{DAC,I}$ indicates the internal DAC in the feedforward path and $K_{DAC,G}$ indicates the global DACs connected to the virtual ground of the integrator. Under perfect matching of $H_1 = STF_2$, E_1

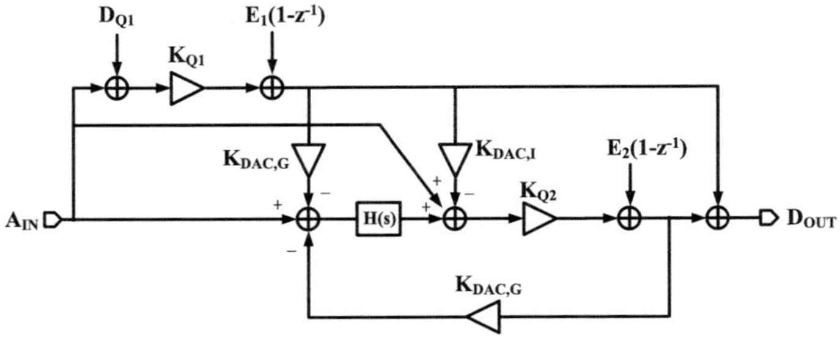

Fig. 7.4 Equivalent signal model of the proposed 0-$\Delta\Sigma$ MASH VCO-based ADC

and D_{Q1} from the 1st-stage quantizer are fully cancelled out at the output, while the quantization noise of the second quantizer is shaped by the analog integrator and the digital differentiator.

7.2.2 *Nonlinearity-Cancellation Robustness Against PVT Variations*

Unfortunately, in practice, due to the process-voltage-temperature (PVT) variations, it is difficult to satisfy the matching condition of $H_1 = STF_2$, which then degrades the performance of the nonlinearity cancellation.

Observed from the signal model in Fig. 7.4, all the nonlinear harmonics appearing at the final ADC output can be expressed as:

$$\begin{aligned} & D_{Q1}K_{Q1}\left(1 - \frac{K_{DAC,G}K_{Q2}H(s) + K_{DAC,I}K_{Q2}}{1 + K_{DAC,G}K_{Q2}H(s)}\right) \\ =& D_{Q1}K_{Q1}\frac{1 - K_{DAC,I}K_{Q2}}{1 + K_{DAC,G}K_{Q2}H(s)} \end{aligned} \tag{7.2}$$

This shows that the cancellation is mainly dependent on the product of the internal DAC gain and the 2nd-stage quantizer gain, and is insensitive to the gain variation of the first VCO and the global DAC. Significantly, the variation of $K_{DAC,I}$ and K_{Q2} is shaped by the loop filter H(s).

MATLAB behavioral simulations have been performed to compare the performance degradation due to any stage-gain mismatch in the 0–1 MASH and 0–2 MASH structures respectively. The results are shown in Fig. 7.5. In the simulations, both 5-bit/5-bit quantizers clocked at 1.6 GHz and a 40-MHz-bandwidth input are used. It shows that the 0–2 MASH VCO-based structure is much more insensitive to gain mismatch than the 0–1 MASH structure. In addition, since the 0–1 MASH structure merely offers 1st-order noise shaping, only a moderate SNDR can be achieved with a reasonable sampling frequency. The power consumption may increase due to the additional integrator and feedback DAC compared to the 0-1 MASH structure. However, by using a low-power design technique at the circuit level, designers can gain more from the higher resolution and robustness, resulting in a better overall FoM.

Furthermore, to investigate the mismatch effect of different blocks, Fig. 7.6 illustrates the SFDR performance of a 0–2 MASH VCO-based ADC as a function of the normalized gain of the quantizers and DACs. The $\pm 10\%$ gain variations of the first quantizer and the global DAC have no evident effect on the SFDR performance. In contrast, the SFDR degrades with the gain variations of the second quantizer and the internal DAC. Thanks to the 1st-order high-pass shaping, the proposed ADC is still robust and achieves 78-dB SFDR with $\pm 10\%$ gain variations.

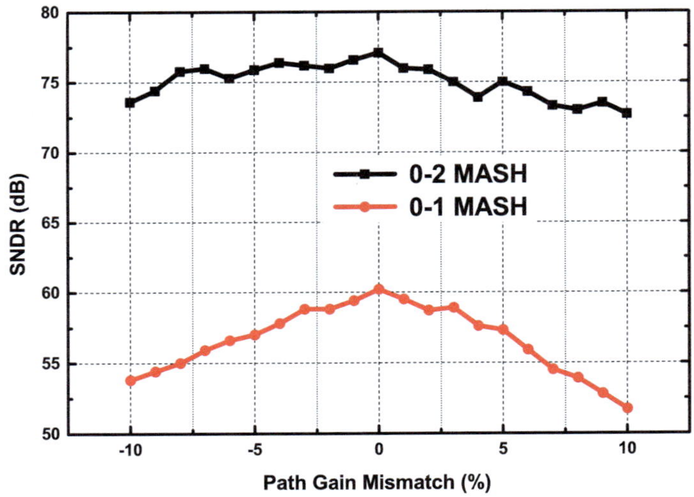

Fig. 7.5 SNDR of the 0–1 MASH and 0–2 MASH VCO-based ADCs as a function of the stage-gain mismatch

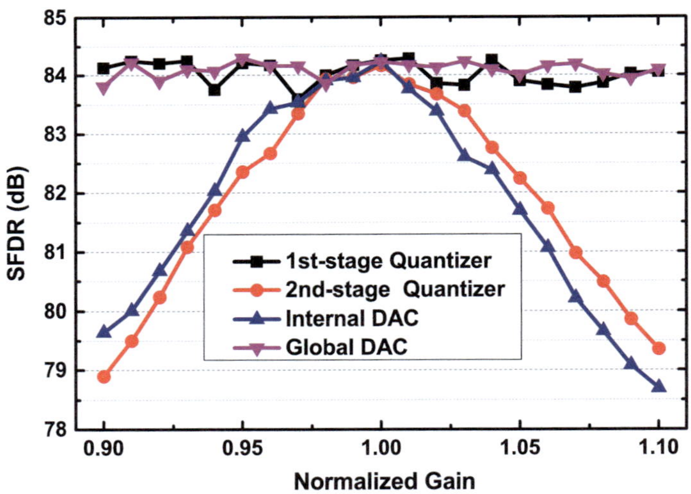

Fig. 7.6 ADC SFDR performance as a function of the normalized gain of the VCOs and DACs

7.2.3 *System Architecture of a 0–2 MASH VCO-Based $\Delta\Sigma$ ADC*

The proposed VCO-based 0–2 MASH $\Delta\Sigma$ ADC is shown in Fig. 7.7. The digital filter H_1 will be illustrated in detail in Fig. 7.10 later on. Nonlinearity exists in different blocks, but it is mitigated in this design at both architecture and circuit levels.

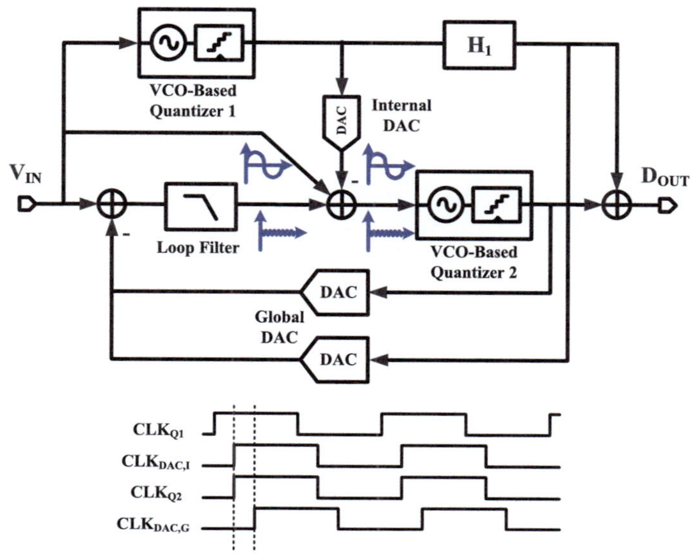

Fig. 7.7 The presented 0–2 MASH VCO-based $\Delta\Sigma$ ADC

Inevitably, the first-stage VCO's nonlinear V-to-F function can introduce harmonics. The low-swing voltage input results in a linear second-stage VCO. As a result, the output of the second quantizer contains harmonics derived from the previous stage rather than from the quantizer itself. The harmonics have the same amplitude levels as the first-stage output but with the opposite phases. When the two stage outputs are added together, these undesired distortions are neutralized. Compared to this design, although the second-stage output in [9] is free from distortion, no cancellation exists to improve the first-stage quantizer's linearity.

Secondly, the integrator opamp only processes quantization noise. In addition, the feedforward path can also reduce the output swing of the opamp. The distortion of the loop filter is therefore not the ADC performance bottleneck.

Finally, the intrinsic DEM in the VCO quantizer for the digital output greatly improves the DAC's linearity. Consequently, considering that the VCO-based quantizer has an inherent 1st-order noise shaping capability, the proposed cascaded VCO-based ADC is equivalent to a linear 2nd-order MASH $\Delta\Sigma$ modulator.

In the practical design implementation, different circuit non-idealities exist and have an impact on the modulator performance. The gain mismatch due to PVT variations has been discussed in Sect. 7.2. As discussed in Chap. 3, clock jitter introduces a sampling error in the VCO-based quantizers. Under perfect matching of $H_1 STF_2$, the jitter-induced sampling noise of the 1st-stage quantizer is cancelled out at the output, while the jitter-induced sampling noise of the 2nd-stage quantizer is reduced by the preceding integrator. However, the clock jitter of the global DACs generates errors which are referred to the input directly. According to the behavioral-level simulations, the distortion-cancellation performance is robust to clock jitter, while the

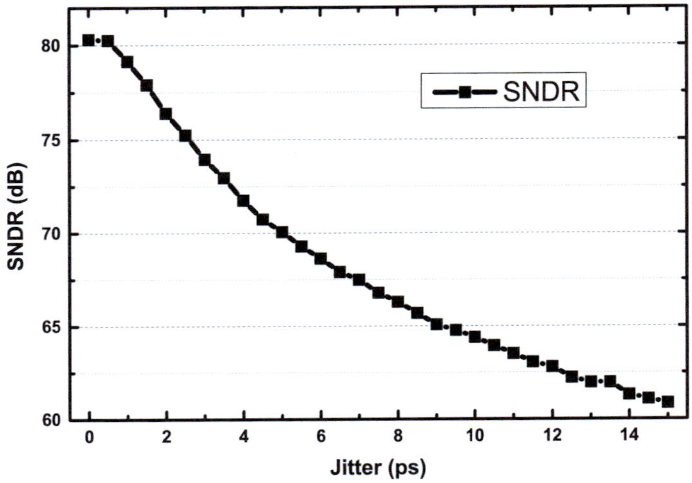

Fig. 7.8 SNDR performance as a function of the clock jitter

ADC noise floor degrades because of the jitter-induced noise in the global DACs. As shown in Fig. 7.8, to reach a 12-bit linearity, the clock jitter must be limited to 4 ps.

7.2.4 *Delay Matching Technique*

The impact of the circuit delay will be investigated in this subsection. As discussed in Sect. 7.2, the nonlinearity cancellation requires $H_1 = STF_2$, which can be simplified to $STF_2 = 1$. However, for a VCO-based quantizer, the VCO's integration operation requires one clock period, meaning that the delay of this quantizer is always larger than the sampling period. Taking this delay into account, STF_2 is no longer flat. The delay matching is therefore crucial for the first-stage nonideality cancellation, especially for wide-bandwidth applications.

Fig. 7.9 Block diagram of a simplified CT loop to approximately represent the 2nd stage in a MASH ADC

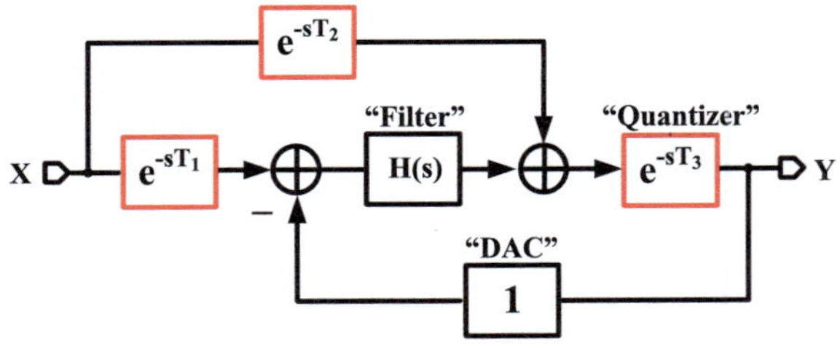

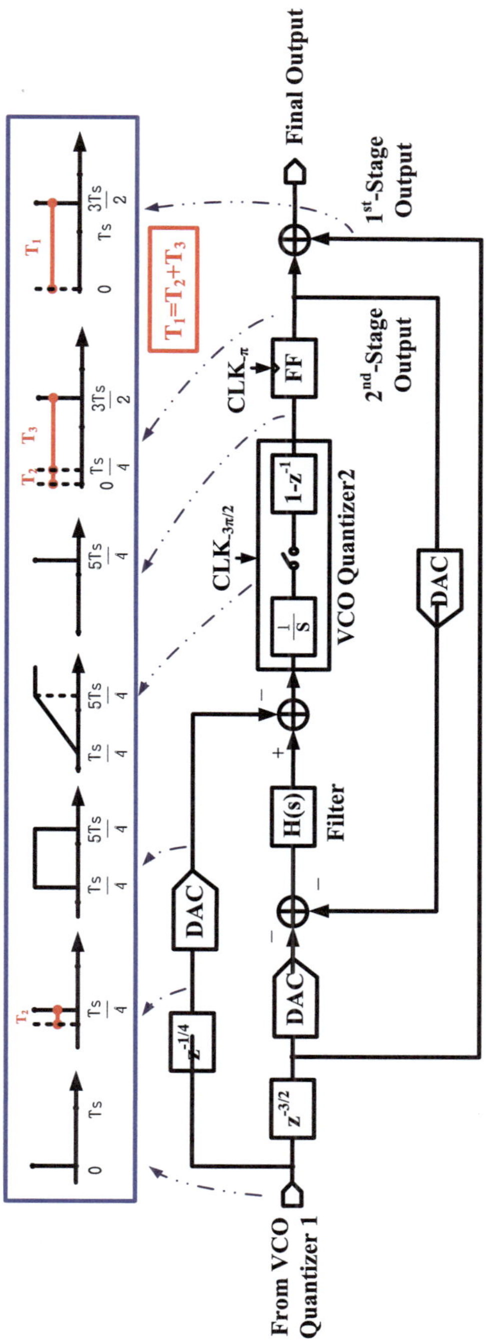

Fig. 7.10 Impulse invariance transformation of the 2nd stage in the presented 0–2 MASH VCO-based ADC

To improve the delay matching, Fig. 7.9 illustrates a simplified continuous-time (CT) loop with different additional delays which approximately represents the 2nd stage. In the presence of these delays, the transfer function of this loop is obtained as:

$$\frac{Y}{X}(s) = e^{-s(T_2+T_3)} \frac{1 + H(s)e^{-s(T_1-T_2)}}{1 + H(s)e^{-sT_3}} \tag{7.3}$$

Specificly, if $T_1 = T_2 + T_3$ is ensured, this transfer function can be simplified as a delay:

$$\frac{Y}{X}(s) = e^{-sT_1} \tag{7.4}$$

In this case, if we also add the delay e^{-sT_1} to the first-stage output, the cancellation condition is satisfied.

The schematic from the 1st-stage output to the 2nd-stage output is re-drawn in Fig. 7.10. To study the delay matching, an impulse input is applied at the 1st-stage output. The impulse invariance transformation is shown. A quarter clock delay, introduced between the first quantizer and the feedforward internal DAC, is used to absorb the quantizer's propagation delay and to re-synchronize the output signals before entering into the DAC. This clock delay refers to the delay T_2 shown in Fig. 7.9. Similarly, a flip-flop (FF) succeeding the second quantizer determines the delay T_3. Then we can introduce the delay $T_1 = T_2 + T_3 = 1.5T_S$ preceding the second-stage input. Therefore, the delay matching is satisfied as needed to improve the nonlinearity cancellation.

7.3 Circuit Implementation of the 0–2 MASH VCO-Based $\Delta\Sigma$ ADC

7.3.1 Three-Input Adder

As shown in Fig. 7.7, a three-input adder is needed to sum up all the branches preceding the second quantizer. Normally, two kinds of adder, active and passive, can be employed. To reduce the power consumption, it is preferred to remove the power-hungry active components from the adder. Compared to the active adder, a passive adder is preferred as it has no power-hungry active components and thus no extra DC power is wasted. However, the summed signal is also attenuated, which is not suitable for the nonlinearity cancellation.

In this work, a novel dual-input VCO-based quantizer is proposed in the 2nd stage to solve this design challenge. As shown in Fig. 7.11, a two-input passive adder combined with an implicit current adder realizes the summation function. In this circuit, an inherent current adder is implemented in the dual-input VCO structure, where the sum of the two current-based signals represents the adder output,

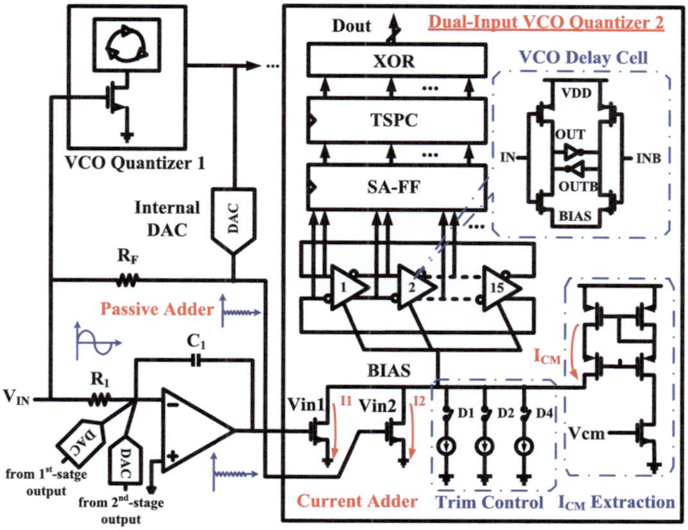

Fig. 7.11 Topology of the dual-input VCO-based fine quantizer (single-ended) to realize the three-input adder

introducing no penalty in terms of DC power or latency. The extra common-mode (CM) current is subtracted through a cascode current mirror. Thanks to the shaping from the preceding loop filter, this three-input adder does not require a 12-bit accuracy in this design. Since both voltage swings of the two VCO inputs are small enough, the proposed dual-input VCO-based quantizer does not cause any nonlinearity problem.

7.3.2 *VCO-Based Quantizer*

The schematic of the VCO-based quantizer [10] is also shown in Fig. 7.11. To cancel out the even-order harmonic distortion [11], the pseudo-differential 5-bit quantizers clocked at 1.6 GHz are composed of two 4-bit single-ended VCO-based quantizers. Each ring VCO contains 15 current-starved delay cells, nominally operating at about 400 MHz, with a VCO gain K_V of about 2 GHz/V. A sense-amplifier-based flip-flop (SAFF) regenerates and samples the converted phase signal. Its schematic is given in Fig. 7.12. At the end, a subsequent true-single-phase clock (TSPC) register and an XOR gate fulfill the 1st-order differentiation function.

The phase noise of the first VCO is also canceled in the cascade structure, while the phase noise of the second VCO can be regarded as 1st-order high-pass shaped noise and is added to the 2nd-order-shaped quantization noise E2. As shown in Fig. 7.13, the simulated phase noise is −115 dBc/Hz at 10-MHz offset frequency. Following the formula in [12] and considering the one extra order of shaping from the analog

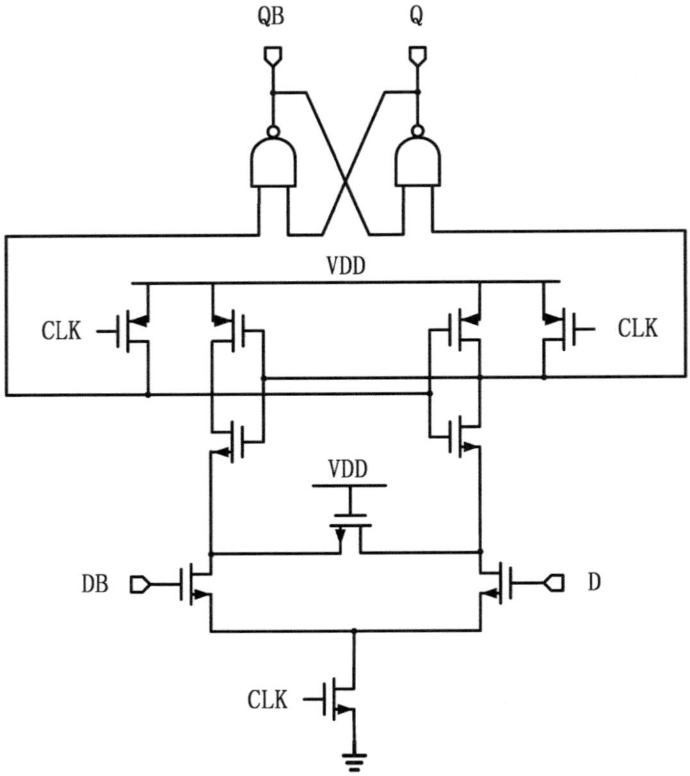

Fig. 7.12 Schematic of the sense-amplifier-based flip-flop

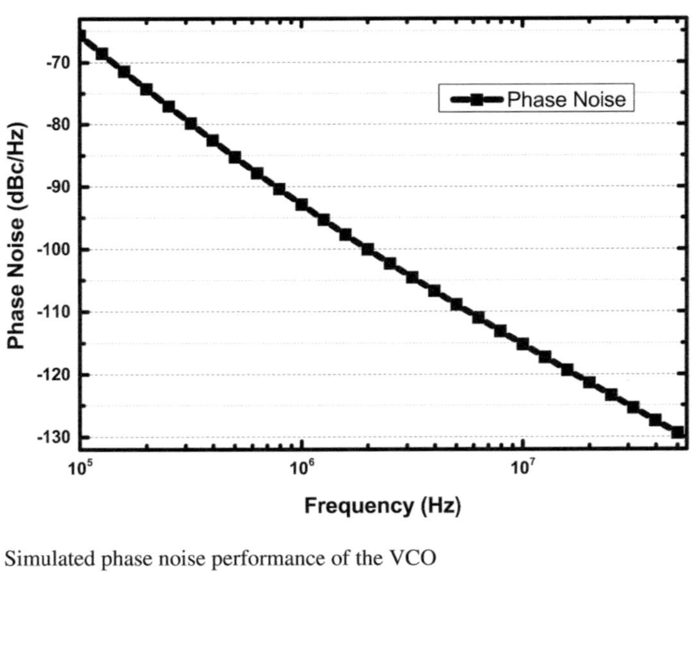

Fig. 7.13 Simulated phase noise performance of the VCO

integrator, the corresponding calculated signal-to-phase-noise-ratio is 102 dB. It is clear that this phase noise does not limit the ADC performance.

For PVT variations, simulations show that the power-supply sensitivity of the VCO gain is 13.6%/V while the temperature sensitivity is 2.2/°C. Considering the operation conditions in practice (e.g., ±20% power supply voltage variation and 50-°C temperature range), the maximum allowed ±10% gain accuracy requirement of the second VCO-based quantizer is sufficient for the 12-bit ADC accuracy in this design. In addition, 3-bit binary-controlled current-trimming cells are utilized to tune the VCO's biasing current, therefore the impact of process variations can be accommodated.

7.3.3 Integrator

The structure of the analog integrator used is shown in Fig. 7.14. Due to PVT variations, the RC time constant can vary as much as ±30%. To accmodate this effect, a 4-bit binary-weighted capacitor array, shown in Fig. 7.15, is used to tune the time constant. The total equivalent capacitance is:

$$C_{total} = 8C + kC, k = 0, 1, ..., 15 \tag{7.5}$$

The minimum and maximum available capacitor values are $C_{min} = 8C$ and $C_{max} = 23C$. So the tuning range of the tunable capacitor is:

$$\frac{C_{max}}{C_{min}} = 2.88 \tag{7.6}$$

and the tuning resolution is given by:

$$\frac{C}{16C} = 6.25\% \tag{7.7}$$

where the 16C is the nominal value of the capacitor bank, which can accomodate the effect of PVT variations.

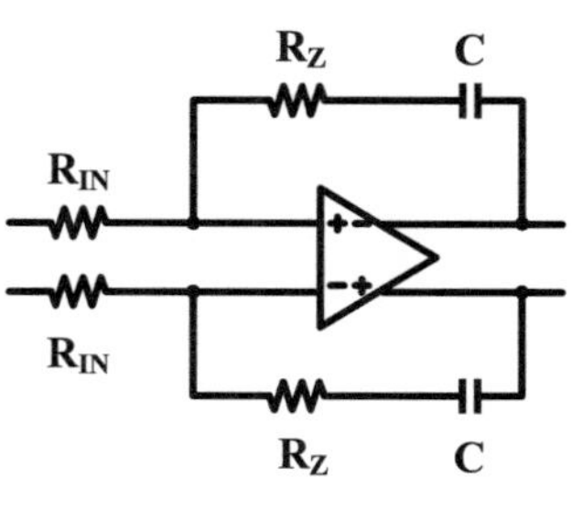

Fig. 7.14 Schematic of the integrator

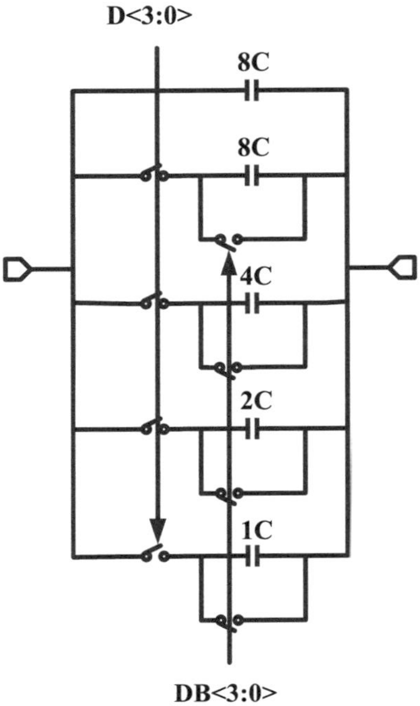

Fig. 7.15 Discretely tunable capacitor for the loop filter

In the layout, shielding is carefully placed at some critical locations, e.g., the virtual ground, to avoid unwanted crosstalk with high-frequency digital signals. In the measurements, an off-line manual capacitance tuning targeted at the 6.25% tuning resolution is adopted. After tuning, the capacitance value remains fixed during the ongoing measurements. An RC time constant auto-tuning structure might be implemented in the future. It only needs to detect the noise floor of the second stage output since the first stage is open loop.

The operational amplifier used in the integrator is implemented with a fully-differential topology as shown in Fig. 7.16. In order to save power consumption, the Miller compensation scheme is eliminated. Instead, a left-half-plane zero introduced by a feed-forward path is used to compensate the phase shift. The simplified small-signal equivalent block diagram of this amplifier is illustrated in Fig. 7.17. The transfer function of the amplifier is given by:

$$H(s) = (g_{m1}R_1g_{m2}R_2 + g_{m3}R_2) \cdot \frac{1 + s\frac{R_1C_1g_{m3}R_2}{g_{m1}R_1g_{m2}R_2+g_{m3}R_2}}{(1 + sR_1C_1)(1 + sR_2C_2)} \tag{7.8}$$

Clearly, a left-half-plane zero is generated as:

$$z \approx -\frac{g_{m1}g_{m2}}{g_{m3}C_1} \tag{7.9}$$

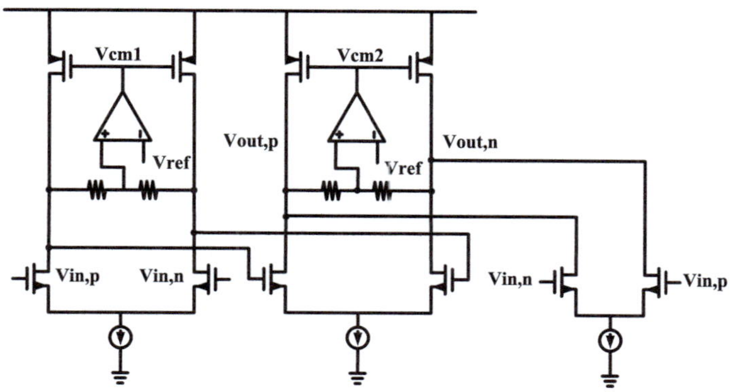

Fig. 7.16 Schematic of the operational amplifier in the integrator

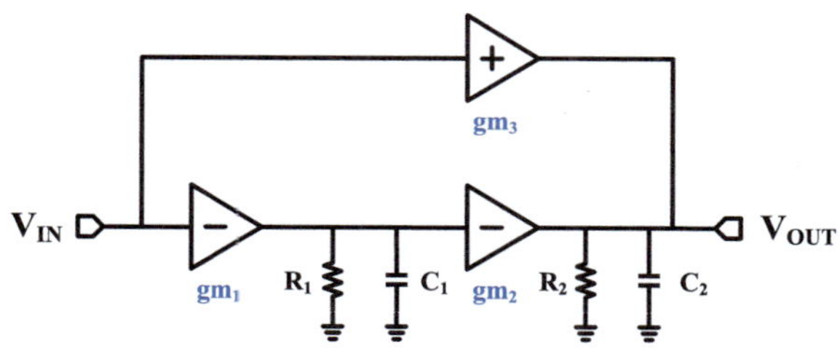

Fig. 7.17 Small-signal equivalent block diagram of the operational amplifier

The DC levels at the two-stage outputs are controlled by two common-mode feedback (CMFB) blocks. Opamp with high DC gain are usually required in a traditional MASH structure to maintain $NTF_1 = H_2$ and to reduce the noise leakage. However, as discussed in Sect. 7.3, thanks to the open-loop first stage, the matching requirement is greatly relaxed in this novel structure. In other words, the noise leakage is mitigated even with a finite-gain opamp. Transistor-level simulations predict that 31-dB DC gain and 1-GHz GBW are achieved with 1.5-mW DC power and 1.5-pF load, which satisfies the requirements needed according to the MATLAB behavioral simulations.

7.3.4 DACs

According to the behavioral simulations in Matlab, Fig. 7.18 illustrates the impact of the DAC cells mismatch. The inherent DEM provided by the VCO-based quantizer reduces the linearity degradation due to the DAC cells mismatch. To achieve a 12-bit accuracy (74-dB SNDR), the standard deviation of the global DACs needs to be limited to 8%, while the matching requirement of the internal DAC is greatly relaxed because it is further shaped by the analog integrator. The explicit DEM is therefore avoided, saving power and area.

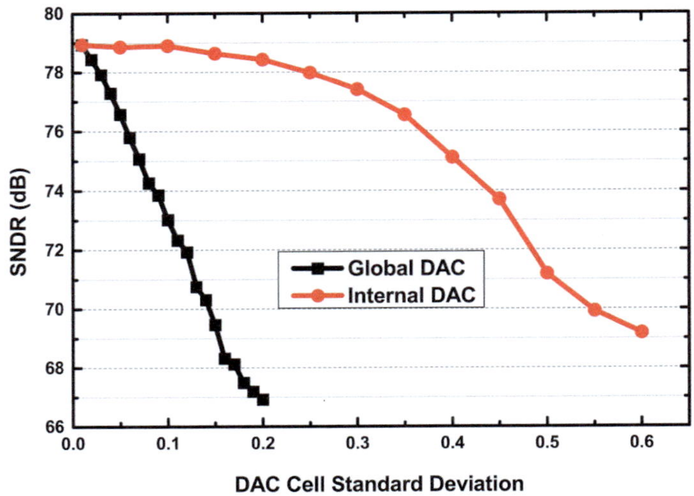

Fig. 7.18 ADC SNDR performance as a function of the standard deviation of the current of the global and internal DAC cells

The DACs in the main feedback path and in the internal feedback path are implemented using a non-return-to-zero (NRZ) current-steering structure. The global current-feedback DACs are the dominant noise sources in this design. The current-source transistors are designed with large dimensions and high overdrive voltage to reduce the noise contribution. Noise simulations show that the root-mean-square (RMS) output noise voltage of the global DAC is 36 μV, which corresponds to about -78 dBFS.

7.3.5 *Interface*

The output of the single-ended quantizer is a 4-bit thermometer-coded signal. A unary-to-binary encoder, as shown in Fig. 7.19, is integrated on chip to transfer the output codes from unary to binary type, and then to sum them up. The function of the encoder is to count the number of ones in the thermometer-coded input. A Wallace tree topology is adopted considering the trade-off of speed and power.

The implemented mirror adder in the encoder is presented in Fig. 7.20. The NMOS and PMOS chains are completely symmetrical. A maximum of two series transistors appear in the carry generation path.

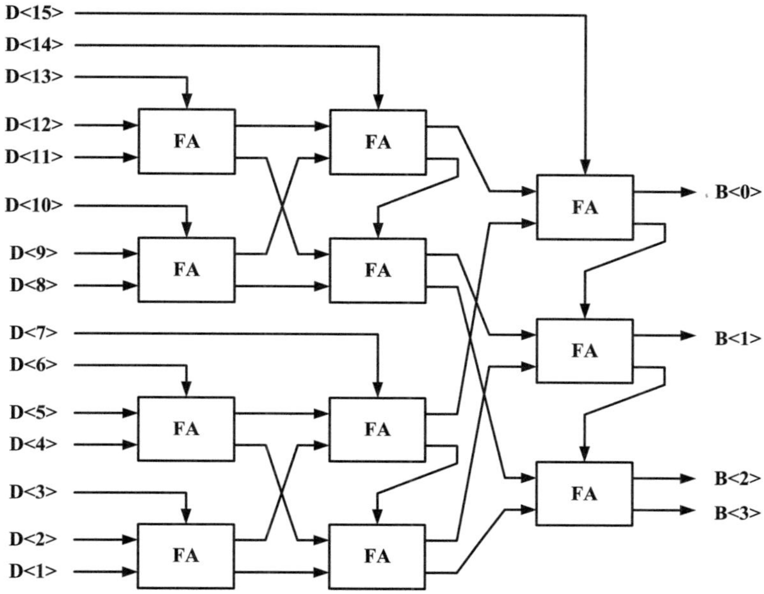

Fig. 7.19 Block diagram of the Wallace tree encoder

7.4 Experimental Results

The prototype design of the presented 0–2 MASH VCO-based ADC has been fabricated in a 40-nm CMOS process with 0.9-V supply voltage. The chip microphotograph is shown in Fig. 7.21. The core circuit occupies an active area of 0.16 mm^2. The modulator output bitstreams are fed to digital buffers and to a low-voltage differential signal (LVDS) interface. As the main purpose of this chip is to verify the proposed architecture and to investigate the critical design considerations, the decimation filter has not been implemented in this design. A multi-stage configuration might be adopted for this in future work.

The measurement setup is shown in Fig. 7.22. A microcontroller board is implemented to program the on-chip shift register which can adjust the digital-controlled bits of the biasing current bank in the VCOs and the capacitor bank in the integrator.

The averaged 8-length 16384-point measured PSD is shown in Fig. 7.23, with an 8-MHz −2.5-dBFS 0.5-V_{PP} input signal. The top of Fig. 7.23 presents the output of the 1st stage. A strong 3rd-order harmonic with 1st-order noise shaping is observed. The signal level in the middle of Fig. 7.23 is about 20 dB lower, providing that the 2nd-stage quantizer only processes the residual signal. Its harmonic has the same amplitude level as for the first quantizer. Finally, when adding the two paths together,

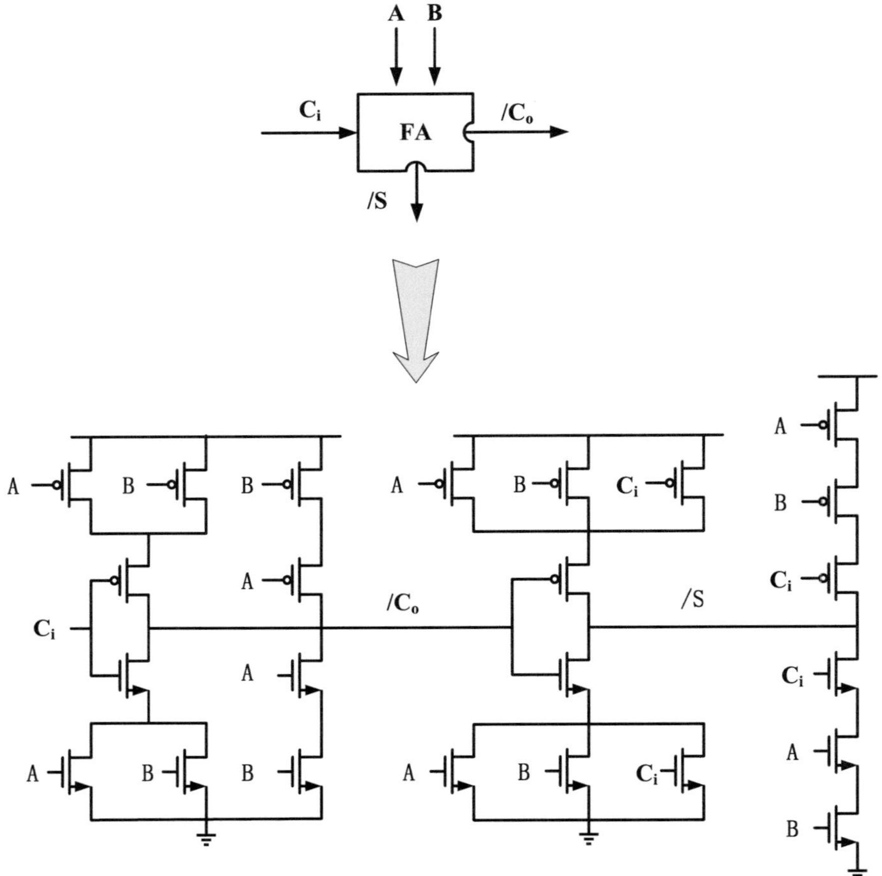

Fig. 7.20 Schematic of the mirror adder implemented in the unary-to-binary encoder

as shown in the bottom of Fig. 7.23, the distortion is largely cancelled and a 77.1-dB SFDR is obtained for the complete ADC. Figures 7.24 and 7.25 present the measured spectrum with a 4-MHz and 1-MHz input respectively.

The measured SNDR performance as a function of the input level for different input frequencies is shown in Fig. 7.26. The peak SNDR for a 1-MHz, 4-MHz and 8-MHz input are 68.9 dB, 67.8 dB and 66.8 dB, respectively. As indicated in Eq. (7.2), the distortion at the final output is frequency-dependent due to the high-pass shaping. Therefore the ADC prototype with a lower-frequency input achieves a higher peak SNDR. The 8-MHz frequency of the input signal represents the worst case because it is the highest frequency for which the 5th-order harmonics fall within the 40-MHz

BW. In addition, a feedforward path is added in the second stage to maintain STF_2 flat at high frequencies. Therefore, the ADC performance can be maintained over the whole signal bandwidth. With an 8-MHz input, the ADC achieves a peak SNR/SNDR of 68.7/66.8 dB over a 40-MHz signal bandwidth, resulting in an ENOB of 10.81 bit.

A two-tone intermodulation test has been conducted with a 10-MHz and 11-MHz input. The signal levels are both −8.2 dBFS. In Fig. 7.27, it is shown again that the IM3 is frequency-dependent because of the 1st-order high-pass shaping. The low-frequency 3rd-order IMD located at 9 and 12 MHz are −90 dBFS, while the high-frequency 3rd-order IMD at 31 and 32 MHz are about 8 dB higher. The corresponding IM3 is 73 dB, which is nearly the worst case because the high-frequency 3rd-order IMD falls within the 40-MHz BW.

The total power consumption is 4.98 mW under 0.9-V supply, resulting in an excellent FoM of 165.8 dB (SNDR+10log(BW/Power)) or 35 fJ/step (Power/ (2BW*2^{ENOB})). The performance summary of the 0–2 MASH VCO-based $\Delta\Sigma$ ADC is shown in Table 7.1, and will be compared to the state-of-the-art works at the end of Chap. 8, after having presented the final ADC design. The presented design achieves the best FoM among all the state-of-art wide-bandwidth (>20 MHz) $\Delta\Sigma$ ADCs, as far as we know.

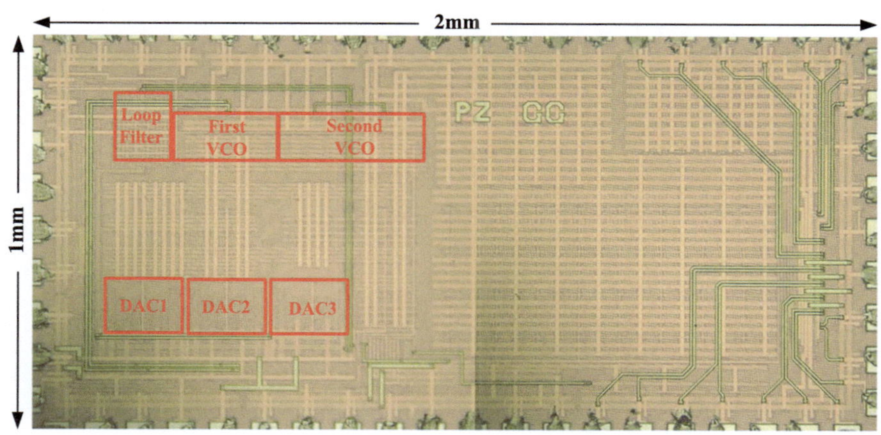

Fig. 7.21 Chip microphotograph of the 0–2 MASH $\Delta\Sigma$ ADC prototype

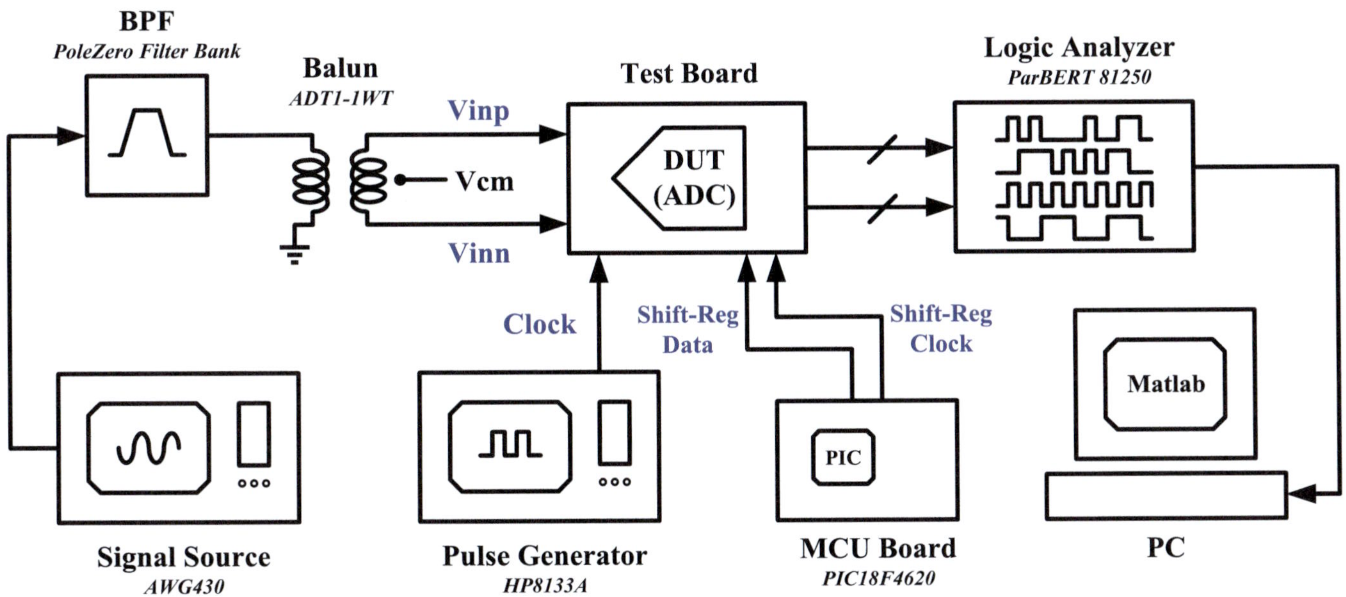

Fig. 7.22 Measurement setup for the 0–2 MASH $\Delta\Sigma$ ADC prototype

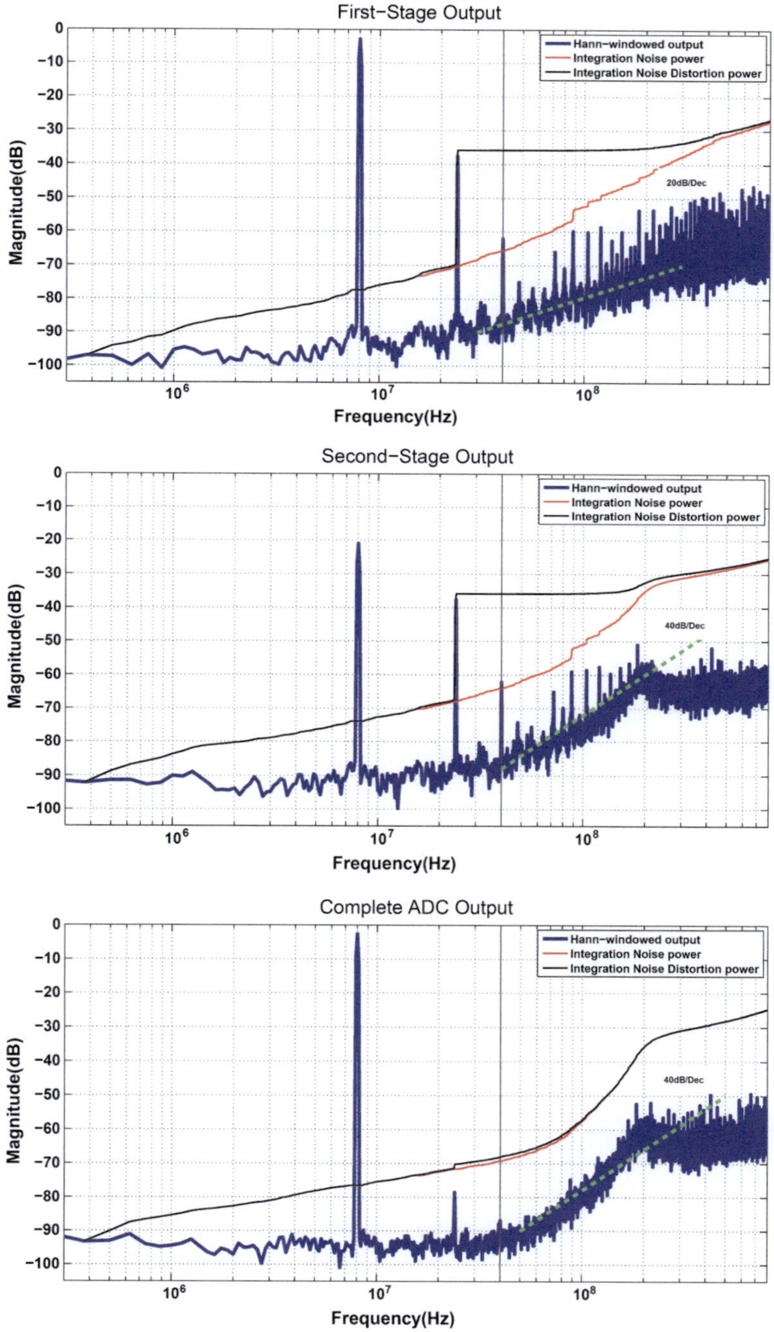

Fig. 7.23 Measured PSD results of the 0–2 MASH $\Delta\Sigma$ ADC chip prototype with an 8-MHz input. *Top* first stage; *middle* second stage; *bottom* complete ADC output

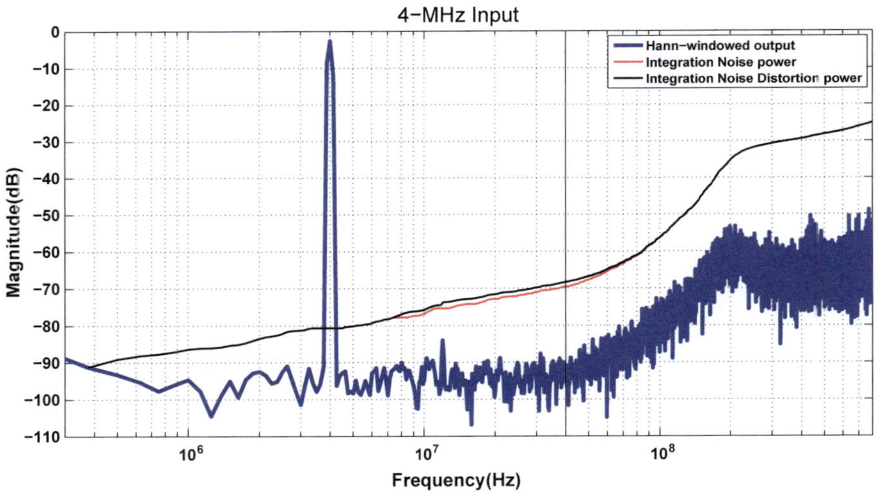

Fig. 7.24 Measured PSD results of the 0–2 MASH $\Delta\Sigma$ ADC chip prototype with a 4-MHz input

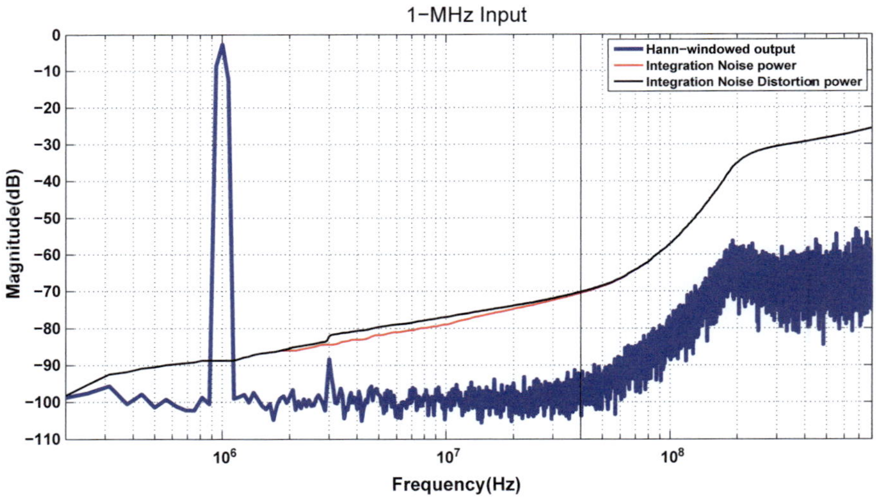

Fig. 7.25 Measured PSD results of the 0–2 MASH $\Delta\Sigma$ ADC chip prototype with a 1-MHz input

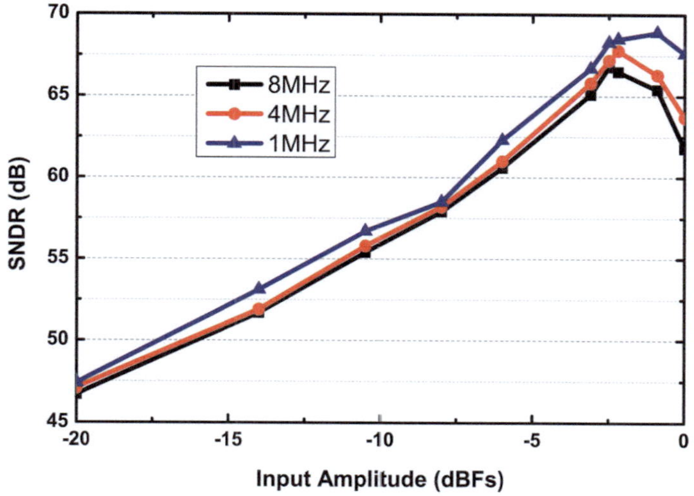

Fig. 7.26 Measured SNDR versus the input amplitude for different input frequencies

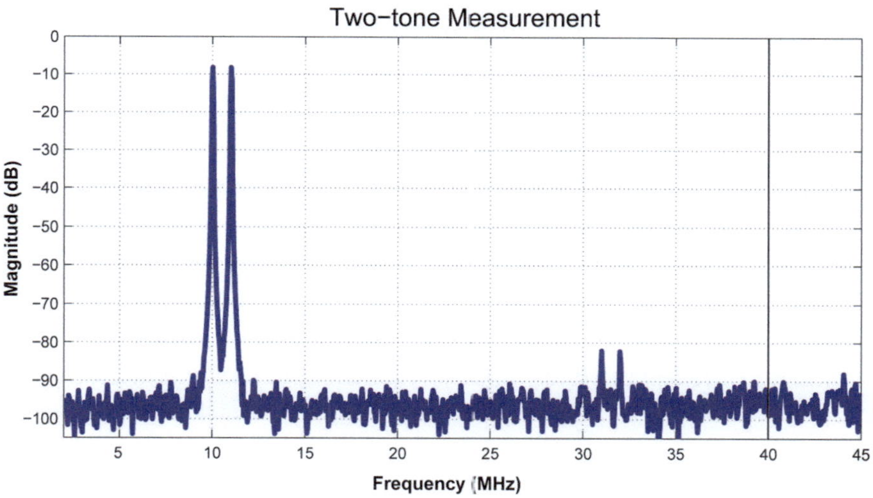

Fig. 7.27 Output spectrum of the two-tone measurement (at 10 and 11 MHz)

Table 7.1 Measurement summary of the 0–2 MASH VCO-based $\Delta\Sigma$ ADC

Specification	Value
Signal bandwidth	40 MHz
Sampling frequency	1.6 GHz
Peak SNR	68.7 dB
Peak SNDR	66.8 dB
Power	4.98 mW
FoM_1	35 fJ/Step
FoM_2	165.8 dB
Core area	0.16 mm^2
Technology	TSMC 40 nm GP

7.5 Conclusions

VCO nonlinearity is the bottleneck of VCO-based $\Delta\Sigma$ ADCs with moderate accuracies. This chapter has presented a nonlinearity-cancellation technique in a power-efficient 0-$\Delta\Sigma$ MASH ADC structure with dual-input VCO-based ADC. A 0–2 MASH VCO-based $\Delta\Sigma$ ADC prototype implemented in 40-nm CMOS confirms the theoretical analysis and achieves a SNDR of 66.8 dB over a 40-MHz BW, consuming only 4.98 mW. By taking full advantage of the low-power VCO-based circuit and overcoming its inherent nonlinearity problem, the FoM of the state-of-art high-bandwidth $\Delta\Sigma$ ADCs has been extended to 166 dB or 35 fJ/step. Compared with the design presented in Chap. 6, this 0–2 MASH VCO-based ADC design has achieved better accuracy performance and even better power efficiency, an analog integrator is still needed however. To implement a highly-digital architecture and high-order noise shaping simultaneously, a fully-VCO-based 0–2 MASH ADC will be shown in the next Chapter, without any analog amplifier and with an excellent FoM.

References

1. P. Zhu, X. Xing, G. Gielen, A 40MHz-BW 35fJ/step-FoM nonlinearity-cancelling two-step ADC with dual-input VCO-based quantizer, in *IEEE European Solid-State Circuits Conference (ESSCIRC), 2014*, (Venice, Italy, 2014), pp. 63–66
2. P. Zhu, X. Xing, G. Gielen, A 40-MHz bandwidth 0–2 MASH VCO-based Delta-Sigma ADC with 35-fJ/step FoM. IEEE Trans. Circuits Syst. II Express Briefs **62**(10), 952–956 (2015)
3. A. Gharbiya, D.A. Johns, A 12-bit 3.125 MHz bandwidth 0–3 MASH Delta-Sigma modulator. Solid-State Circuits, IEEE J. **44**(7), 2010–2018 (2009)
4. Y. Dong, R. Schreier, W. Yang, S. Korrapati, A. Sheikholeslami, A 235 mw CT 0–3 MASH ADC achieving −67 dbBFS/Hz NSD with 53 MHz BW, in *Solid-State Circuits Conference Digest of Technical Papers (ISSCC), 2014 IEEE International*, (IEEE, 2014), pp. 480–481
5. A. Sanyal, K. Ragab, L. Chen, T. Viswanathan, S. Yan, and N. Sun, A hybrid SAR-VCO Delta-Sigma ADC with first-order noise shaping, in *2011 IEEE Symposium on CICC*, (2014), pp. 1–4

6. X. Xing, G.G.E. Gielen, A 42 fJ/step-FoM two-step VCO-based Delta-Sigma ADC in 40 nm CMOS. Solid-State Circuits, IEEE J. **50**(3), 714–723 (2015)
7. K. Reddy, S. Rao, R. Inti, B. Young, A. Elshazly, M. Talegaonkar, P.K. Hanumolu, A 16-mW 78-dB SNDR 10-MHz BW CT $\Delta\Sigma$ ADC using residue-cancelling VCO-based quantizer. Solid-State Circuits, IEEE J. **47**(12), 2916–2927 (2012)
8. A. Gupta, K. Nagaraj, T. Viswanathan, A two-stage ADC architecture with VCO-based second stage. IEEE Trans. Circuits Syst. II Express Briefs **58**(11), 734–738 (2011)
9. W. Yu, J. Kim, K. Kin, S. Cho, A time-domain higher-order MASH Delta-Sigma ADC using voltage-controlled gated-ring oscillator. IEEE Trans. Circuits Syst. I: Regul. Pap. **59**(8), 856–866 (2012)
10. M.Z. Straayer, M.H. Perrott, A 12-bits, 10-MHz bandwidth, continuous-time Delta-Sigma ADC with a 5-bits, 950-MS/s VCO-based quantizer. IEEE J. Solid-State Circuits **43**(4), 805–814 (2008)
11. S. Rao, K. Reddy, B. Young, P. Hanumolu, A deterministic digital background calibration technique for VCO-based ADCs. IEEE J. Solid-State Circuits **49**(4), 950–960 (2014)
12. J. Kim, T.K. Jang, Y.G. Yoon, S.H. Cho, Analysis and design of voltage-controlled oscillator based analog-to-digital converter. IEEE Trans. Circuits Syst. I Regul. Pap. **57**(1), 18–30 (2010)

Chapter 8
Fully-VCO-Based High-Order $\Delta\Sigma$ ADC

8.1 Introduction

Scaling friendly in the light of the reducing voltage headroom and increasing time resolution of the nanometer CMOS technology, VCO-based quantizers have been exploited in CT $\Delta\Sigma$ ADCs. As discussed in Chap. 7, the 0-$\Delta\Sigma$ MASH topology overcomes the VCO nonlinearity problem due to its inherent operation principle, being more power-efficient than a single-loop structure. The analog integrator in the $\Delta\Sigma$ loop, however, remains difficult to be implemented with technology scaling and consumes a large amount of the power as a high OTA DC gain is required. Conceptually, a VCO accumulates the excess phase like an ideal integrator. It is therefore beneficial to shift the analog integrator to the time and the digital domains as well.

In this chapter, we will report a 0-2 MASH CT $\Delta\Sigma$ ADC with fully-VCO-based integrator and quantizers, without any analog integrators. Firstly the principle of a VCO-based integrator will be presented in Sect. 8.2. The circuit design details and silicon implementation results will be given and discussed in Sect. 8.3. Taking advantage of the fully-digital high-order structure and relieving the VCO nonlinearity bottleneck, a 74-dB SFDR and a FoM of 162 dB or 52 fJ/step over a 40-MHz bandwidth are achieved. Finally, the conclusion will be drawn in Sect. 8.4.

8.2 Integrators

8.2.1 *Traditional Analog Integrator*

In a traditional $\Delta\Sigma$ ADC, the first integrator of the loop filter needs to satisfy a severe gain and speed requirement due to the absence of the noise shaping, therefore it consumes most of the power. Compared to the amplifers implemented in switched-

X. Xing et al., *Design of Power-Efficient Highly Digital Analog-to-Digital Converters for Next-Generation Wireless Communication Systems*, Signals and Communication Technology, https://doi.org/10.1007/978-3-319-66565-8_8

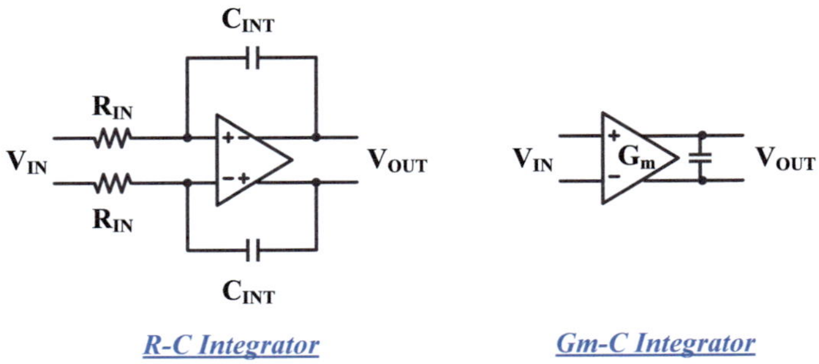

Fig. 8.1 Traditional analog R-C and G_m-C integrator

Fig. 8.2 Simplified diagram of a R-C integrator, in which the opamp has a finite DC gain and a finite GBW

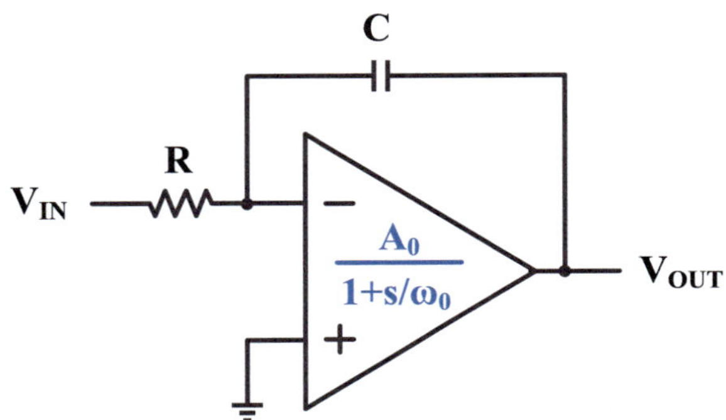

capacitor (SC)-based DT $\Delta\Sigma$ ADCs, the opamps in CT $\Delta\Sigma$ ADCs have a relaxed speed requirement. However, the non-idealities of the integrator, such as the finite GBW and finite DC gain, also degrade the modulator performance.

Conventionally, as shown in Fig. 8.1, the integrators are implemented by R-C or G_m-C structures. Compared to the active R-C integrator, the open-loop G_m-C structure consumes less power, but has a worse linearity performance.

As an example, a simplified diagram of an R-C integrator, in which the opamp has a finite DC gain A_0 and a finite bandwidth ω_0, is shown in Fig. 8.2. The transfer function of the integrator is obtained as:

$$H(s) \approx -\frac{A_0}{1 + s(A_0 RC + \frac{1}{\omega_0}) + \frac{RC}{\omega_0} s^2} \tag{8.1}$$

Assuming that the DC gain is high enough to satisfy $A_0 \gg 1$, the above transfer function can be simplified as:

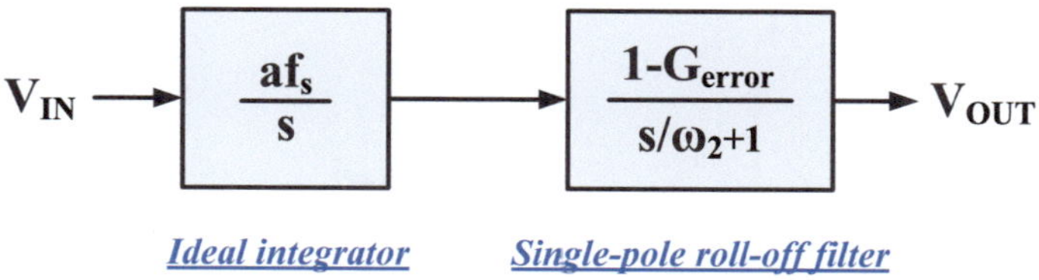

Fig. 8.3 Equivalent model of an R-C integrator, in which the opamp has a finite gain and GBW

$$
\begin{aligned}
H(s) &\approx -\frac{1}{s(RC+\frac{1}{GBW_0})+\frac{RC}{GBW_0}s^2} \\
&= -\frac{1}{sRC}\cdot\frac{1}{1+\frac{1}{GBW_0RC}+\frac{1}{GBW_0}s} \\
&= -\frac{1}{sRC}\cdot\frac{1-G_{error}}{1+\frac{s}{\omega_2}}
\end{aligned} \tag{8.2}
$$

where GWB_0 is the GBW of the opamp. G_{error} and ω_2 refer to the gain error and an additional pole caused by the opamp's finite GBW, which are defined by:

$$G_{error} = \frac{\frac{1}{RC}}{GBW_0+\frac{1}{RC}} \tag{8.3}$$

and

$$\omega_2 = GBW_0 + \frac{1}{RC} \tag{8.4}$$

Therefore, the finite GBW results in both a gain error and a time delay. The gain error can be compensated by tuning the integrator coefficient. However, the additional delay leads to a decreased noise suppression, or even loop instability. As shown in Fig. 8.3, the transfer function of an R-C integrator is modeled as an ideal integrator along with a 1st-order roll-off filter. Moreover, the finite DC gain also results in a reduced noise shaping ability of the integrator. Therefore, to increase the DC gain and the GBW especially in a nanometer CMOS process, a penalty of considerable power consumption needs to be paid.

8.2.2 VCO-Based Integrator

To take advantage of the technology scaling that has greatly decreased the power dissipation of the digital parts, it is beneficial to shift the analog integrator to the digital domain as well. As discussed in Chap. 3, a VCO acts like an integrator for the output phase. The frequency-domain transfer function is depicted in Fig. 8.4. A pole

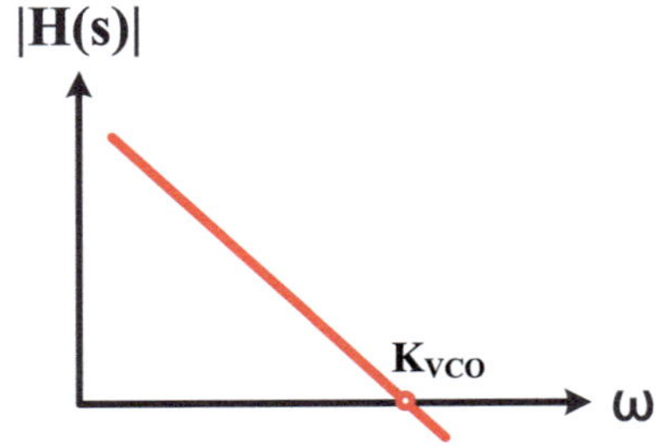

Fig. 8.4 Frequency-domain transfer function of a VCO

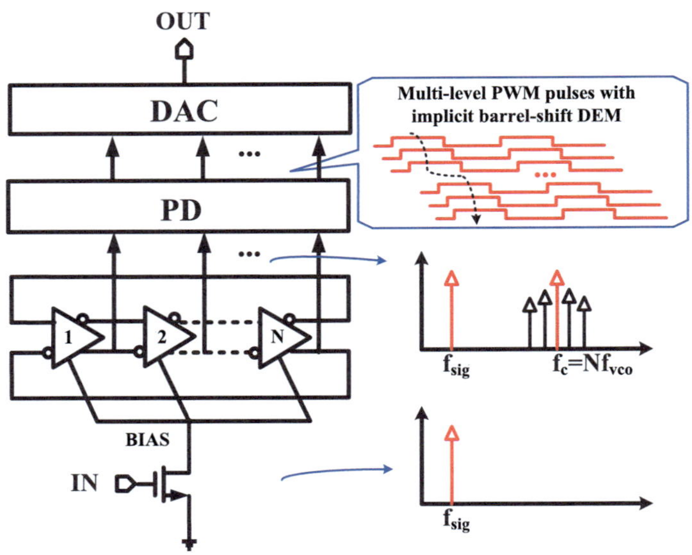

Fig. 8.5 Block diagram of a N-stage VCO-based 1st-order integrator

located at DC is provided with a unity gain frequency K_{VCO}. Clearly, an advantage of a VCO-based integrator is that a VCO theoretically has an infinite DC gain which is power-efficient and independent of the supply voltage.

As can be seen in Fig. 8.5, in the frequency domain, the N-stage VCO modulates the input f_{sig} with a carrier frequency f_C which is defined by:

$$f_C = N \cdot f_{VCO} \tag{8.5}$$

where f_{VCO} is the free-running frequency of the VCO. The VCO output phases are converted to duty-cycled PWM pulses by phase detectors (PD), and then transferred back to voltage/current form by DACs. Interestingly, the barrel-shifting property of the multi-level PWM signals offers an implicit DEM of the DAC cells.

In summary, the total transfer function of this integrator H_{INT} is given by:

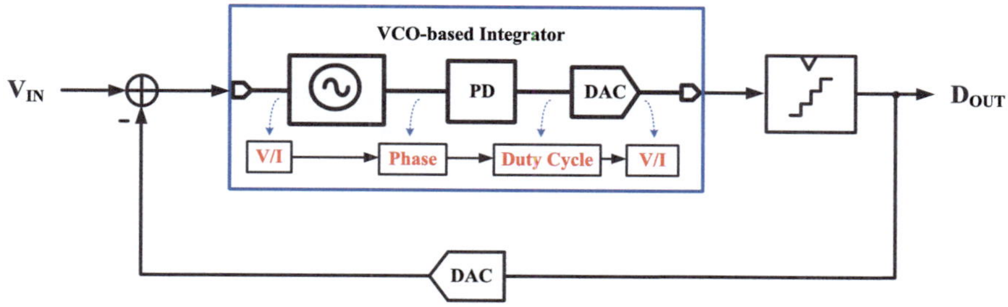

Fig. 8.6 Block diagram of a single-loop $\Delta\Sigma$ ADC with a VCO-based integrator

$$H_{INT}(s) = \frac{K_{VCO} \cdot K_{PD} \cdot K_{DAC}}{s} \tag{8.6}$$

where K_{VCO}, K_{PD} and K_{DAC} are the gain of the VCO, PD and DAC respectively. Moreover, the VCO-based integrator can be extended to a higher order by cascading the basic structures. Reference [1] presents a silicon prototype of a VCO-based 4th-order Butterworth filter by cascading two biquads.

However, as discussed in Chap. 4, the VCO-based integrator suffers from the VCO nonlinearity. The techniques presented in Chap. 4 can be used to mitigate this nonideality. A design example will be shown in Sect. 8.3.

8.2.3 Fully-VCO-Based $\Delta\Sigma$ ADC Structure

Replacing the conventional analog integrator by a VCO, the VCO-based integrator can be utilized in a $\Delta\Sigma$ loop. Figure 8.6 illustrates the block diagram. For the VCO-based integrator, since the sampling operation does not take place here, no quantization error is added. However, f_C needs to be sufficiently high to avoid spurs folding back into the baseband due to the later sampling. The VCO input swing is reduced due to the feedback and the integration. The VCO nonlinearity is therefore mitigated.

As shown in Fig. 8.7, an architecture of a VCO-based $\Delta\Sigma$ ADC with a VCO-based integrator is proposed in [2], with a 2nd-order noise shaping coming from the VCOs in the integrator and in the quantizer. This purely-VCO-based ADC is power-efficient and scaling friendly, benefiting from the mostly-digital structures. However, the ADC performance is still limited by the nonlinearity of the VCO-based quantizer.

The 0-$\Delta\Sigma$ MASH topology discussed in Chap. 7 can overcome the VCO's nonlinearity problem due to its inherent operation principle, being more power-efficient than the single-loop structure. By replacing the analog integrator, the architecture of a fully-VCO-based 0-$\Delta\Sigma$ MASH ADC is proposed in Fig. 8.8. This design will now be elaborated in detail.

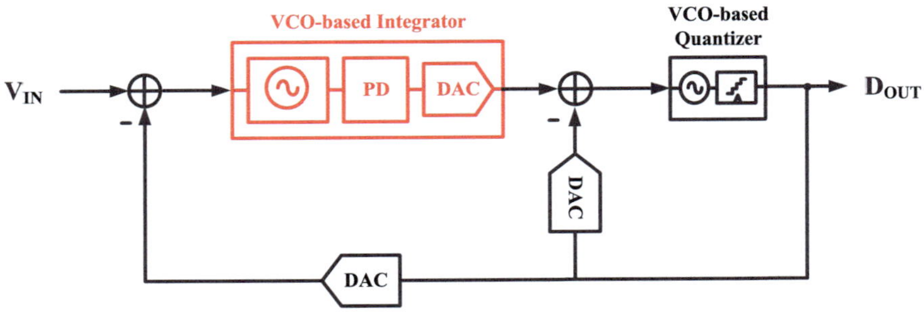

Fig. 8.7 Block diagram of a 2nd-order VCO-based $\Delta\Sigma$ ADC with a VCO-based integrator [2]

Fig. 8.8 Block diagram of the proposed 0-$\Delta\Sigma$ MASH ADC with VCO-based integrators

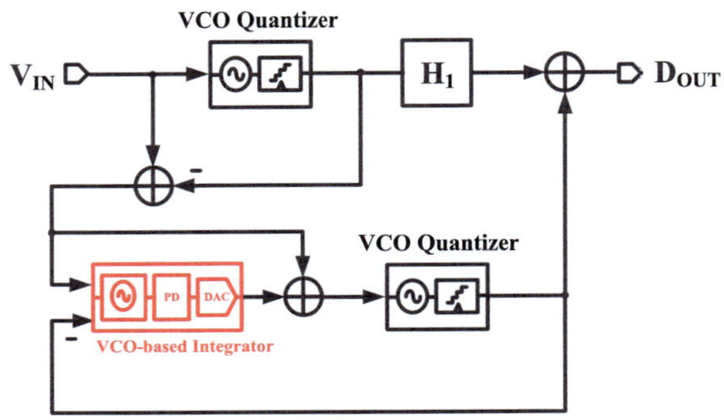

8.3 Design Example: A Fully-VCO-Based 0-2 MASH VCO-Based $\Delta\Sigma$ ADC

8.3.1 System Architecture

A design example of a fully-VCO-based 0-2 MASH VCO-based $\Delta\Sigma$ ADC is presented in this section. Figure 8.9 shows the block diagram of the proposed two-stage ADC. According to the analysis in Chap. 7, the nonlinearity of the first stage is cancelled if the STF of the 2nd stage is equal to 1. An input-feedforward structure is adopted in the second stage to improve the cancellation. As the second stage only processes the quantization noise of the first stage, its input voltage swing is greatly reduced. Therefore, the VCOs implemented in the integrator and in the quantizer of the second stage do not limit the ADC nonlinearity performance.

The clock timing of the proposed ADC is shown in Fig. 8.10. The details concerning delay matching have been discussed in Chap. 7 and [3].

Figure 8.11 depicts the presented MASH ADC and its corresponding signal model, where D_{Q1} represents all the nonlinear harmonics generated in the first quantizer due to the VCO's nonlinear behavior, E_1 and E_2 represent the corresponding quantization noise from the two quantizers, K_{Q1} and K_{Q2} indicate the quantizer gains, $K_{DAC,I}$ indicates the internal DAC in the feedforward path and $K_{DAC,G}$ indicates the global

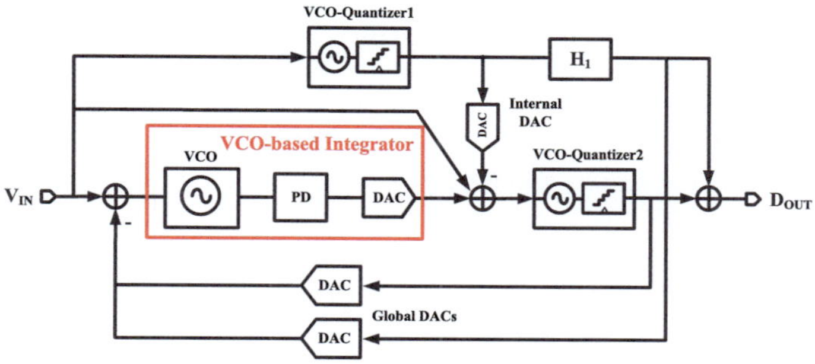

Fig. 8.9 Block diagram of the presented fully-VCO-based 0-2 MASH $\Delta\Sigma$ ADC

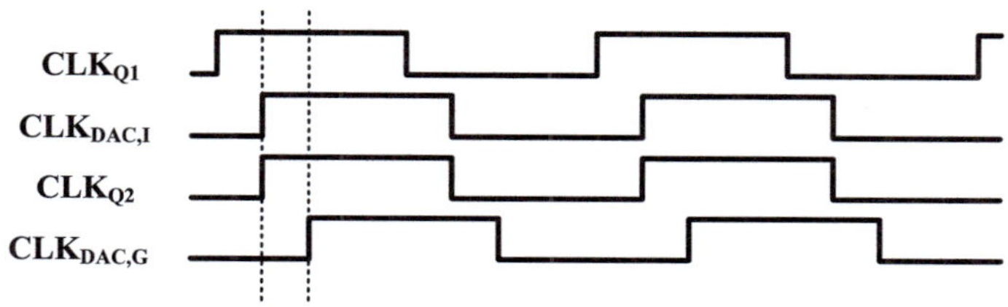

Fig. 8.10 Clock timing of the presented fully-VCO-based ADC

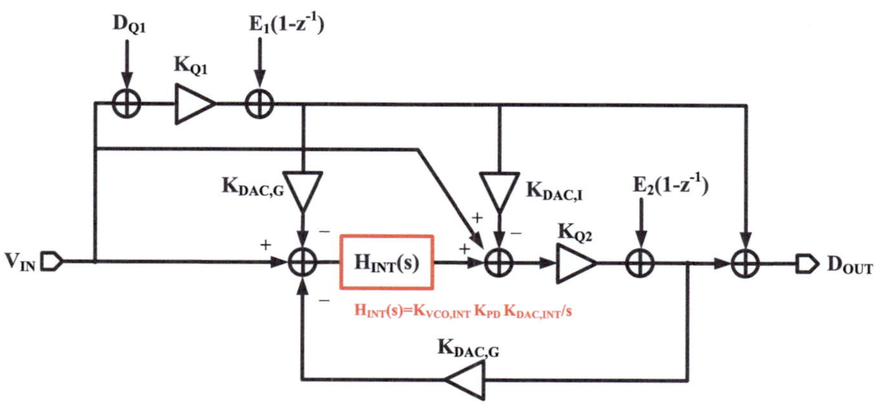

Fig. 8.11 Signal model of the presented fully-VCO-based 0-2 MASH VCO-based ADC

DACs connected to the virtual ground of the integrator. $H_{INT}(s)$ represents the transfer function of the VCO-based integrator, which is defined as Eq. (8.6).

As derived from the signal model, all the nonlinear harmonics appearing at the final ADC output can be expressed as:

$$D_{Q1}K_{Q1}\frac{1 - K_{DAC,I} \cdot K_{Q2}}{1 + \frac{K_{DAC,G} \cdot K_{Q2} \cdot K_{VCO} \cdot K_{PD} \cdot K_{DAC,INT}}{s}} \tag{8.7}$$

This shows that the nonlinearity cancellation is mainly dependent on the product of $K_{DAC,I}K_{Q2}$, which is shaped by the integrator $H_{INT}(s)$.

In behavioral simulations, the proposed ADC achieves 78-dB SFDR within $\pm 10\%$ gain variations of $K_{DAC,I}K_{Q2}$ and 81-dB SFDR within $\pm 10\%$ gain variations of $K_{VCO,INT}$. Simulations show that the power-supply sensitivity and the temperature sensitivity of the VCO's gain both in the integrator and in the quantizers are robust enough for the targeted 12-bit ADC accuracy in this design. In addition, 3-bit binary-controlled current-trimming cells are utilized to accommodate the impact of process variations.

8.3.2 Circuit Implementation

As shown in Fig. 8.9, the VCO-based integrator transfers the input signal to the phase domain, and then recovers it back to the voltage domain after completing the integration process. The pseudo-differential integrator's schematic is presented in Fig. 8.12, consisting of inverter-based ring oscillators, phase detectors and DACs.

The ring oscillator modulates the input with a carrier frequency f_C, which needs to be sufficiently higher than the sampling frequency to avoid spurs folding back into the baseband. In this design, an f_C of 3.5 GHz is realized with 24 stages and a 145-MHz oscillation frequency f_{VCO}, and is filtered out by the subsequent VCO-based quantizer due to its inherent anti-aliasing property. 3-bit binary-controlled current-trimming cells are utilized to tune the VCO's biasing current. The current-source transistors are designed with large dimensions and high overdrive voltage to reduce the noise current.

In addition, the phase noise of the VCOs in the front-end integrator also adds directly at the ADC input and therefore the VCOs should be carefully designed. Non-minimum-size transistors are utilized in the delay cells to reduce the flicker noise. As shown in Fig. 8.13, the simulated VCO phase noise is -125 dBc/Hz at a 10-MHz offset frequency, which does not limit the ADC performance according to the simulation. To accomodate the parasitic delay in the VCO, an extra delay compensation path is added. A buffer and a feedforward capacitor C_F couple the negative VCO input to the biasing node which introduces a negative zero [4].

A SA-based buffer shown in Fig. 8.14 isolates the VCO delay cell and recovers its output to rail-to-rail. The subsequent differential phase detector compares the phases of the pseudo-differential oscillators and generates the duty-cycled output, eliminating the external reference signal. The duty cycle is proportional to the phase difference between the inputs. Moreover, in this differential-PD structure, the impact of the free-running frequency mismatch on the common-mode signal can be mitigated.

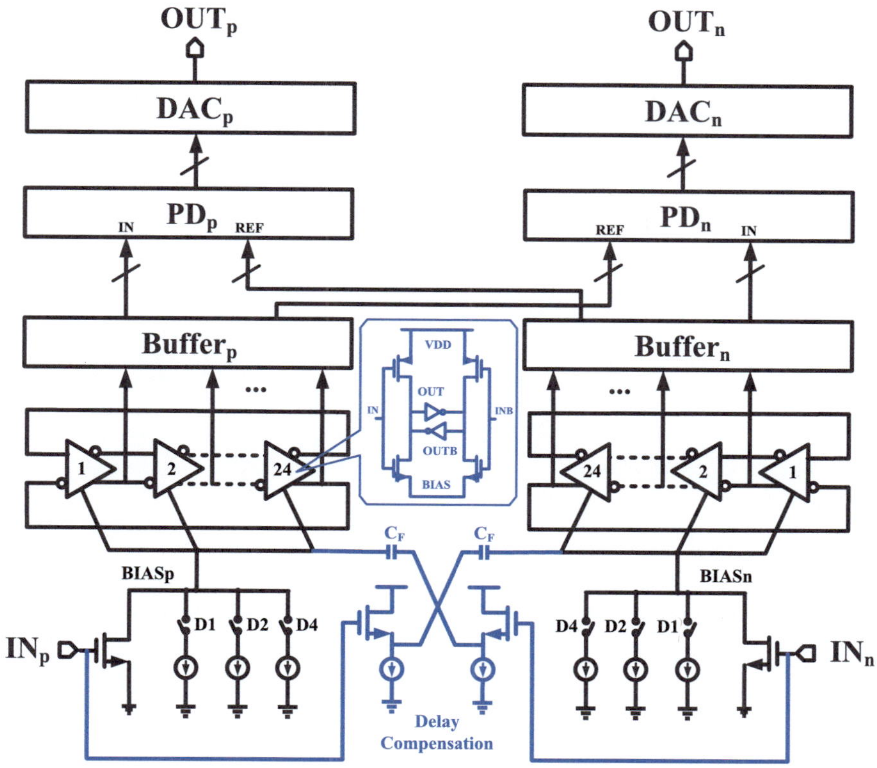

Fig. 8.12 Schematic of the pseudo-differential VCO-based integrator

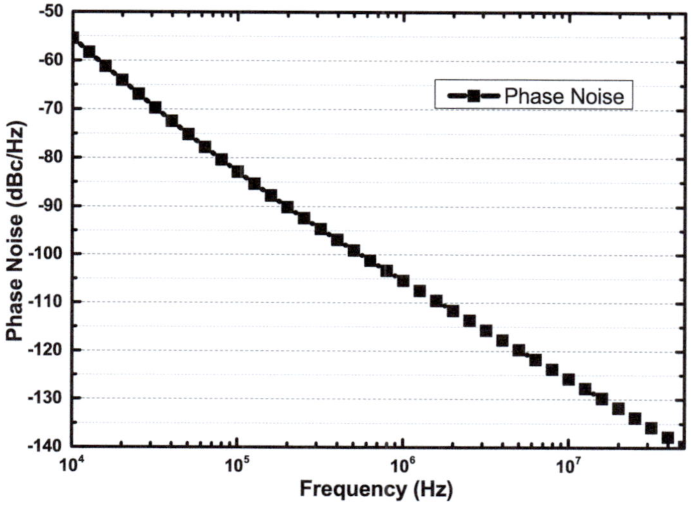

Fig. 8.13 Simulated phase noise performance of the VCO used in the integrator

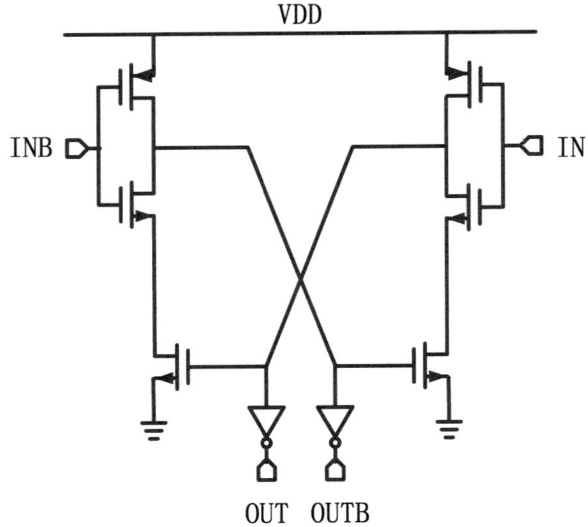

Fig. 8.14 Schematic of the SA-based buffer in the VCO-based integrator

There are three main structures that can be selected as phase detectors:

1. XOR gate as PD:
   The XOR circuit produces error pulses on both the rising and falling edges. When the two inputs are completely in-phase, the output has a constant level of zero. The linear input range is π.
2. Two-state PD:
   The two-state PD generates one error pulse for one comparison. The linear input range is 2π.
3. Three-state phase-frequency detectors (PFD):
   The three-state PFD can produce an output even when the two inputs differ not only in phase but also in frequency. However, the mismatch at the subsequent current subtraction introduces extra nonlinearity [1].

To avoid PD output overflow and to reduce the nonlinearity, the two-state PD is preferred over the XOR gate and the three-state PFD. Therefore, the two-state PD illustrated in Fig. 8.15 is implemented in this design. The transfer function of the two-state PD is shown in Fig. 8.16. The PD outputs a 50%-duty-cycled pulse when the input phase difference is zero, which indicates the common-mode condition. In this design, since the input swing of the second stage is very small, the phase differences always stay in the linear range, and the PD output pulses are duty cycled around 50%. Therefore, the nonlinearity caused by the finite rise/fall time of the PD-generated PWM signals is greatly reduced [1].

As observed from Eq. (8.6), the loop-filter coefficient can be tuned conveniently by adjusting the DAC current. The DACs are implemented using a source-switched structure, as shown in Fig. 8.17. The implicit DEM relaxes the matching requirement of the DAC cells.

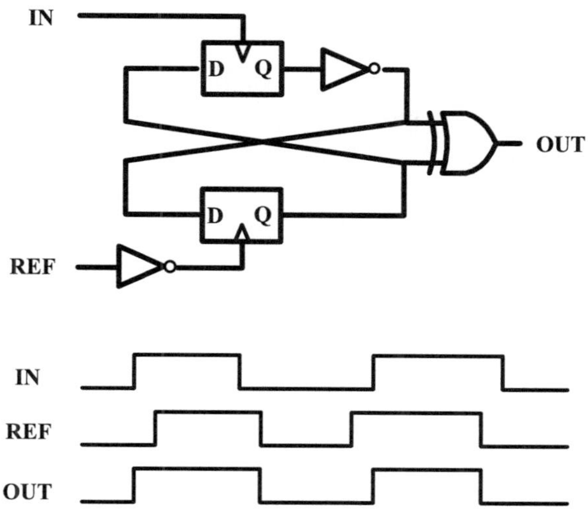

Fig. 8.15 Schematic of the two-state phase detector

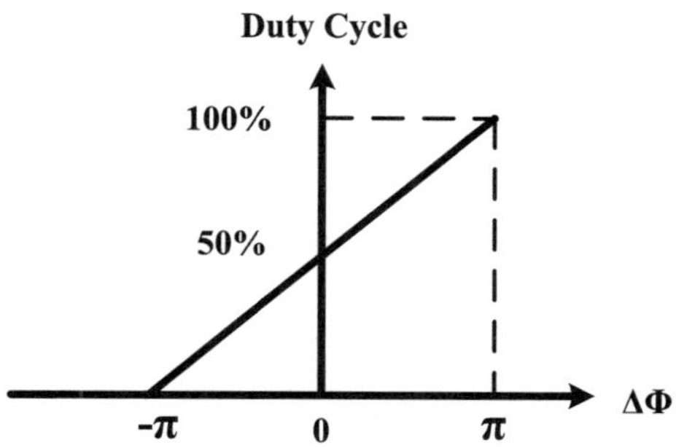

Fig. 8.16 Transfer function of the two-state phase detector

The pseudo-differential 5-bit quantizers in the two stages are clocked at 1.6 GHz. Each single-ended ring VCO contains 15 current-starved delay cells, nominally operating at about 400 MHz. All the details of these building blocks have already been discussed in Chap. 4.

8.3.3 Experimental Results

The prototype design of the presented fully-VCO-based 0-2 MASH $\Delta\Sigma$ ADC has been fabricated in a 40-nm CMOS technology with 0.9-V supply voltage. Figure 8.18 shows the chip micrograph, the core area is 0.17 mm^2. The measurement setup has been shown in the last chapter.

The averaged measured output power spectral density is shown in Fig. 8.19, with an 8-MHz 0.3-V_{P-P} input signal. A strong 3rd-order harmonic exists at the 1st-stage output. When adding the two stages together, the distortion is significantly cancelled and the SFDR is improved from 42 to 74 dB. Folded back by intermodulations,

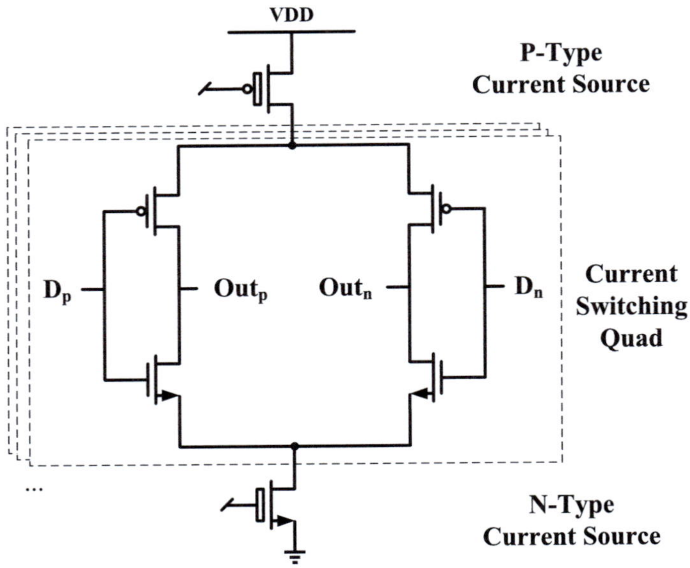

Fig. 8.17 Schematic of the source-switched DAC cells in the VCO-based integrator

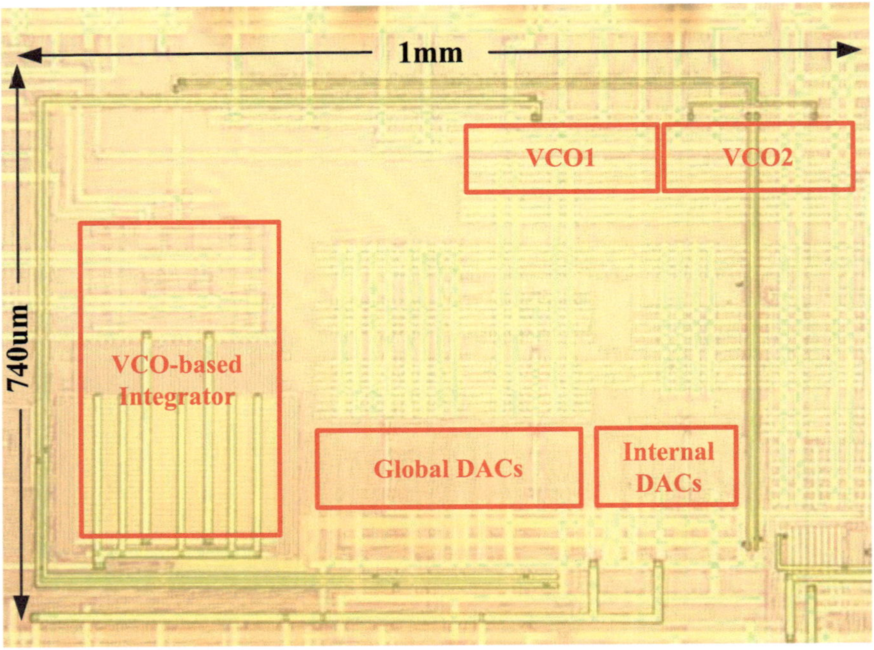

Fig. 8.18 Chip microphotograph of the fully-VCO-based 0-2 MASH $\Delta\Sigma$ ADC prototype

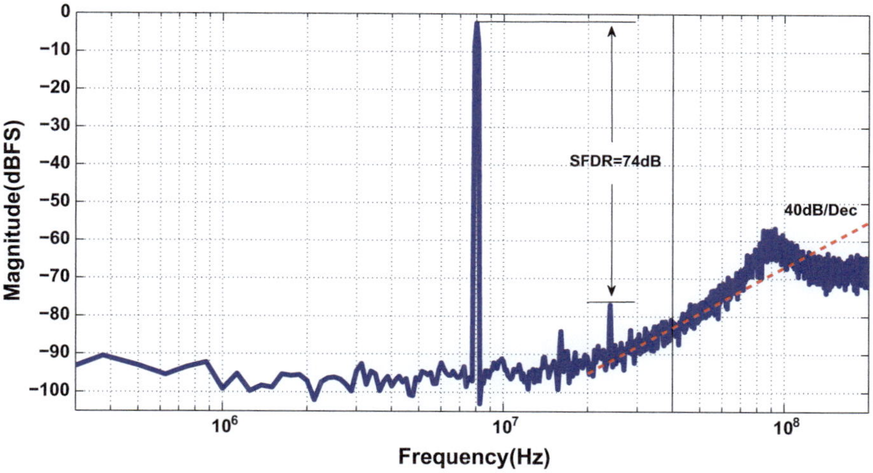

Fig. 8.19 Measured PSD results of the fully-VCO-based 0-2 MASH $\Delta\Sigma$ ADC chip prototype with an 8-MHz input

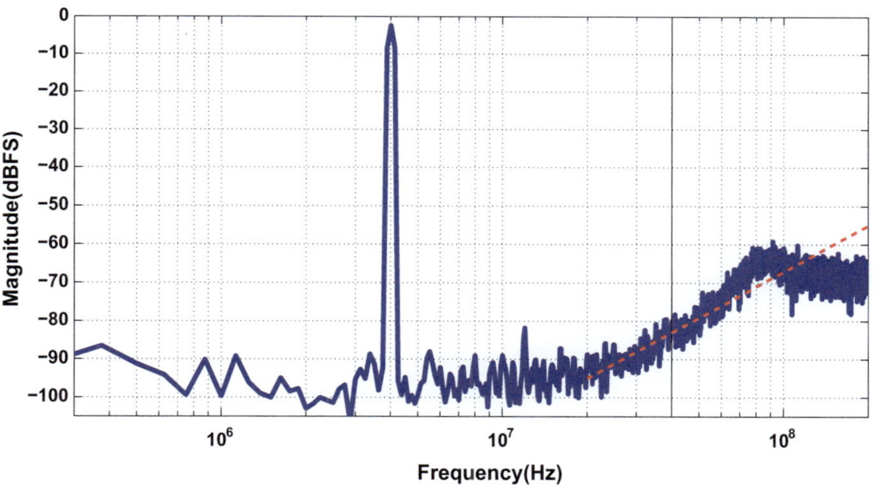

Fig. 8.20 Measured PSD results of the fully-VCO-based 0-2 MASH $\Delta\Sigma$ ADC chip prototype with a 4-MHz input

some spurious tones are observed within the signal band in the PSD plot due to the imperfect cancelation of tones around the oscillation frequency of the multi-level VCOs. A 2nd-order noise shaping is observed, which validates the VCO's integration function. A peak SNR/SNDR of 65.5/62.7 dB is achieved over a 40-MHz signal bandwidth. In comparison, Figs. 8.20 and 8.21 present the measured spectrum with a 4-MHz and 2-MHz input, respectively. The measured SNDR performance as a function of the input level is shown in Fig. 8.22.

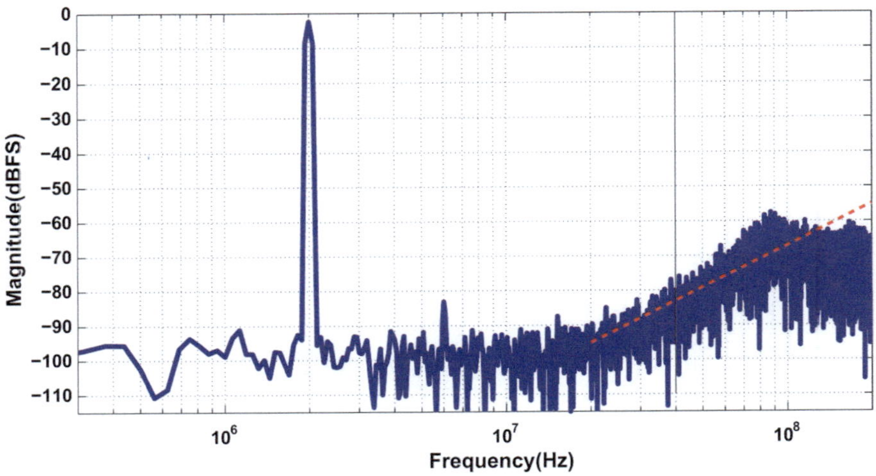

Fig. 8.21 Measured PSD results of the fully-VCO-based 0-2 MASH $\Delta\Sigma$ ADC chip prototype with a 2-MHz input

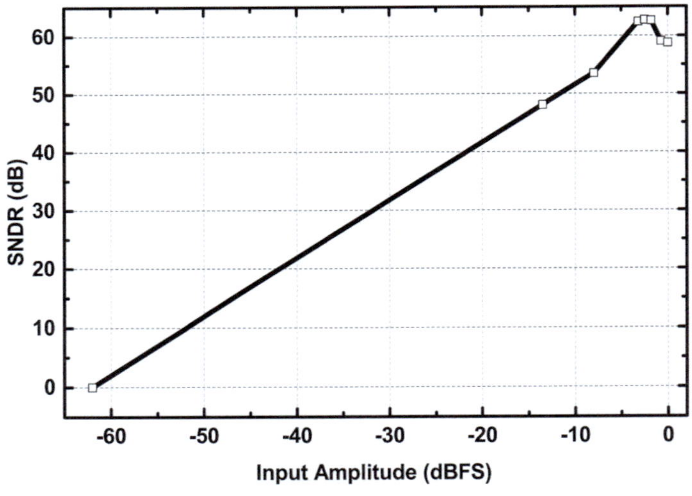

Fig. 8.22 Measured ADC SNDR versus input amplitude

A two-tone intermodulation test has been conducted with a 10 and 11-MHz input. The signal levels are both −8.3 dBFS. In Fig. 8.23, it is shown that the IM3 is frequency dependent because of the 1st-order high-pass shaping. The low-frequency 3rd-order IMD located at 9 and 12 MHz are −86 dBFS, while the high-frequency 3rd-order IMD at 31 and 32 MHz are about 8 dB higher. The corresponding IM3 is 69 dB.

The total power consumption is 4.6 mW, resulting in a FoM of 162 dB (SNDR+10 log(BW/Power)) or 52 fJ/Step (Power/(2BW*2^{ENOB})). This design realizes fully-

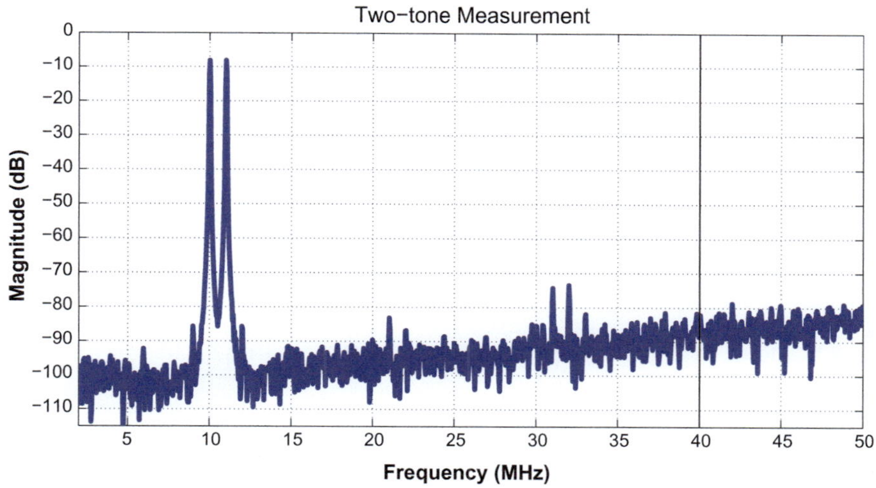

Fig. 8.23 Output spectrum of the two-tone measurement (at 10 and 11 MHz)

Table 8.1 Measurement summary of the fully-VCO-based 0-2 MASH $\Delta\Sigma$ ADC

Specification	Value
Signal bandwidth	40 MHz
Sampling frequency	1.6 GHz
Peak SNR	65.5 dB
Peak SNDR	62.7 dB
Power	4.6 mW
FoM_1	52fJ/Step
FoM_2	162.0 dB
Core area	0.17 mm^2
Technology	TSMC 40 nm GP

VCO-based 2nd-order noise shaping, achieves an excellent FoM and scales more effectively with future CMOS technology.

The measurements summary of the fully-VCO-based 0-2 MASH ADC are shown in Table 8.1. Compared to the design presented in the last chapter which achieves the best FoM among all the state-of-art wide-bandwidth (>20 MHz) $\Delta\Sigma$ ADCs, this fully-VCO-based ADC has a lower power consumption, but with a degraded SNDR performance.

This result is worse than expected. To understand the cause of the linearity degradation, Fig. 8.24 illustrates a digital calibration for the chip prototype, in which a two-tap FIR filter is implemented on the chip's output in Matlab. The additive digital filter, of which the coefficients can be determined through adaptive off-line or on-line calibration [5], aims to reduce the nonlinearity leakage due to the stage-gain mismatch. As shown in Fig. 8.25, after applying this digital correction, the 3rd-order

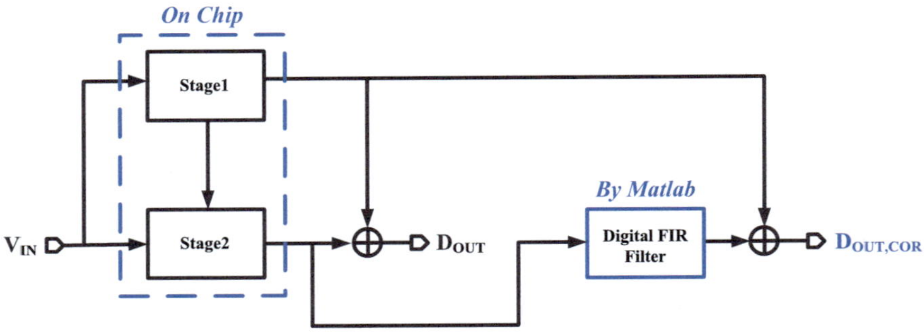

Fig. 8.24 Digital correction setup of the fully-VCO-based 0-2 MASH $\Delta\Sigma$ ADC chip prototype

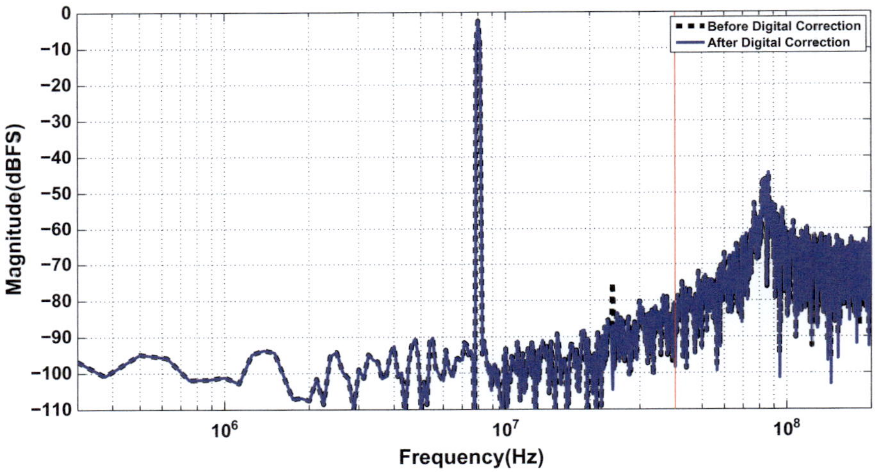

Fig. 8.25 Digitally corrected PSD results of the fully-VCO-based 0-2 MASH $\Delta\Sigma$ ADC chip prototype

harmonic is reduced below the noise floor, while the total harmonic distortion (THD) is improved from 72 to 81 dB, meaning that nonlinearity leakage exists due to PVT variations. Moreover, apart from the nonideal matching, the excess delay resulting form the VCO-based integrator degrades the NTF of the second stage. In future work, local-feedback delay compensation paths can be added in the $\Delta\Sigma$ loop to mitigate this nonideality.

The five ADC designs presented from Chaps. 5–8 are compared to the state-of-the-art oversampling ADCs [6] with BW≥8MHz in Figs. 8.26 and 8.27. In the FoM-BW comparison, as far as we know, the FoM of our 0-2 MASH VCO-based $\Delta\Sigma$ ADC design presented in Chap. 7 is until now the best among all published high-bandwidth (≥20MHz) $\Delta\Sigma$ ADCs, as seen in Fig. 8.27.

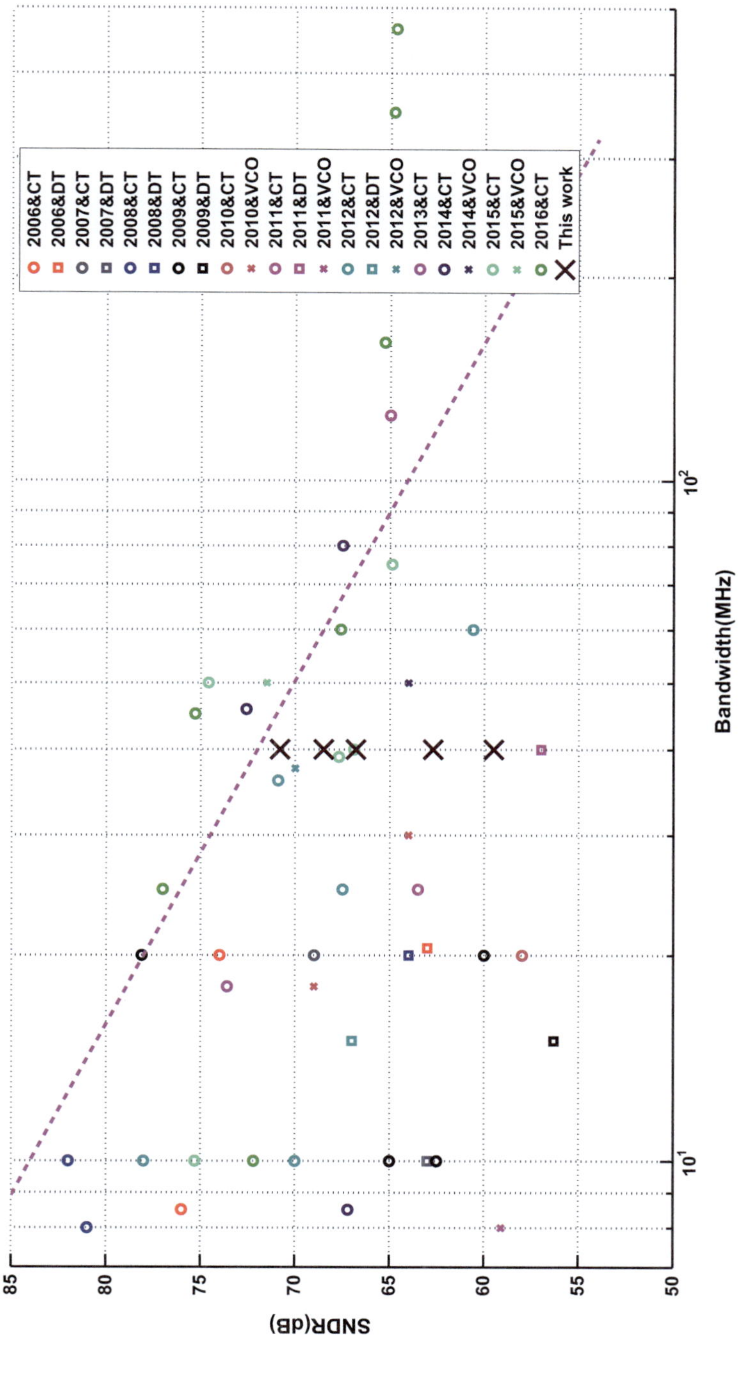

Fig. 8.26 Comparison of the SNDR and the bandwidth of the presented work with the state-of-the-art DSM ADC designs (BW $\geq$8MHz)

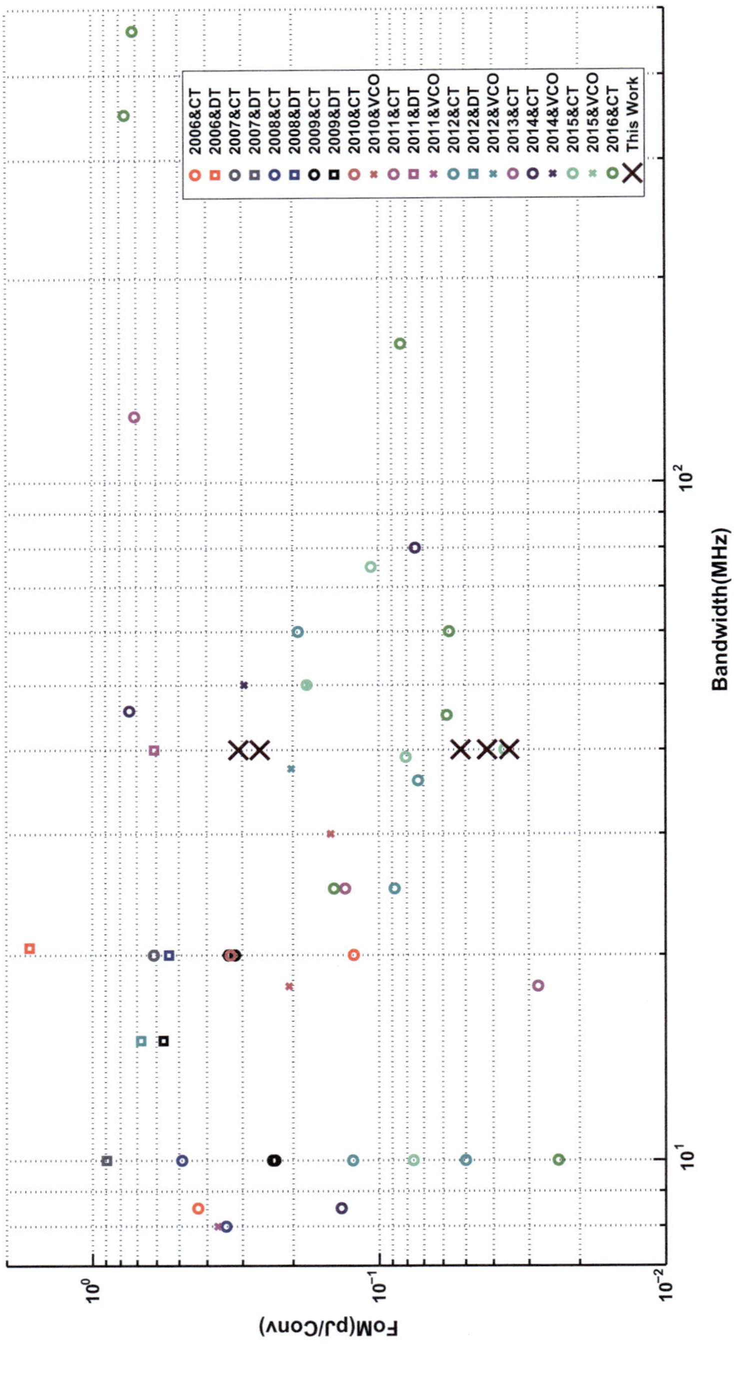

Fig. 8.27 Comparison of the FoM and the bandwidth of the presented work with the state-of-the-art DSM ADC designs (BW ≥8MHz)

8.4 Conclusions

Power consumption is one important design challenge of ADCs. The analog integrator in the $\Delta\Sigma$ loop remains difficult to be implemented with CMOS technology scaling and consumes a large amount of power as a high OTA DC gain is required. Accumulating the excess phase, a VCO-based integrator is more digitally oriented and power efficient. Another advantage is that a VCO theoretically has an infinite DC gain which is power-efficient and independent of the supply voltage.

In this chapter, we have presented a 0-2 MASH CT $\Delta\Sigma$ ADC with fully-VCO-based integrator and quantizers in 40-nm CMOS, without any analog integrators. Benefiting from the 2nd-order noise shaping from the VCOs and relieving the VCO nonlinearity bottleneck by the MASH structure, a SFDR of 74 dB is achieved over a 40-MHz signal bandwidth, with a peak SNR/SNDR of 65.5/62.7 dB. Taking advantage of the fully-digital structure, the power consumption is 4.6 mW, resulting in a FoM of 162 dB or 52 fJ/step. This result is degraded than expected due to the nonlinearity leakage and the excess loop delay.

References

1. B. Drost, M. Talegaonkar, P. Hanumolu, Analog filter design using ring oscillator integrators. IEEE J. Solid State Circuits **47**(12), 3120–3129 (Dec. 2012)
2. Y. Yoon, K. Lee, P. Wang, N. Sun, A purely-VCO-based single-loop high-order continuous-time $\Sigma\Delta$ ADC, in *Proceedings of the IEEE International Symposium on Circuits and Systems (ISCAS)* (June 2014)
3. P. Zhu, X. Xing, G. Gielen, A 40-MHz bandwidth 0–2 MASH VCO-based Delta-Sigma ADC with 35-fJ/step FoM. IEEE Trans. Circuits Syst. II Express Briefs **62**(10), 952–956 (Oct. 2015)
4. K. Reddy, Rao S., R. Inti, B. Young, A. Elshazly, M. Talegaonkar, P.K. Hanumolu, A 16-mW 78-dB SNDR 10-MHz BW CT $\Delta\Sigma$ ADC using residue-cancelling VCO-based quantizer. IEEE J. Solid State Circuits **47**(12), 2916–2927 (Dec. 2012)
5. P. Kiss, J. Silva, A. Wiesbauer, T. Sun, U. Moon, J. Stonick, G. Temes, Adaptive digital correction of analog errors in MASH ADCs i. off-line and blind on-line calibration. IEEE Trans. Circuits Systems II: Analog Digit. Signal Process. **47**(7), 621–628 (Dec. 2010)
6. B. Murmann, ADC performance survey 1997-2016 (2017), http://www.stanford.edu/murmann/adcsurvey.html

Chapter 9
Conclusions

9.1 Summary and Conclusions

The ongoing data explosion requires larger data rates for almost all kinds of communication standards. Another trend for electronic products is that more and more portable devices are powered by a battery. As a result, power consumption is the key design specification. Besides the pressure from the product definition, the receiver architecture and the CMOS technology used also impose several design challenges for low-power ADCs. This book has investigated and demonstrated the design of highly-digital power-efficient DSM ADCs, covering both closed-loop and open-loop architectures. The research work can be summarized as follows.

- In Chap. 2, the basic operation principle of different ADC topologies have shortly been introduced. For each architecture, the typical specifications have been listed and compared. Also, the application of ADC in receiver systems has briefly been presented, and design factors affecting the ADC specifications have been given and discussed.
- Design issues of traditional CT DSMs have been covered in Chap. 3. First the basic techniques of DSM–oversampling and noise shaping–have been introduced. Then different DSM architectures and their advantages/disadvantages have been presented. Finally, important nonidealities of a CT DSM and their modelling have been shown to set the requirements for the design of the building blocks. Corresponding solutions for circuit nonidealities have also been discussed.
- Highly-digital quantization and conversion based on a VCO and digital gates have been presented in Chap. 4. The main design challenge of VCO-based quantizers is the VCO nonlinearity. Several existing closed-loop DSM ADCs and open-loop oversampling ADCs have been introduced that solve the VCO nonlinearity.
- Chapter 5 has discussed the first two designs of the research work, which are 40 MHz-BW 12-bit DSM ADCs in 90 nm CMOS with 4th-order noise-shaping and a 5-bit phase-type VCO-based quantizer. The linearity of the multi-bit feed-

X. Xing et al., *Design of Power-Efficient Highly Digital Analog-to-Digital Converters for Next-Generation Wireless Communication Systems*, Signals and Communication Technology, https://doi.org/10.1007/978-3-319-66565-8_9

back DAC has been solved by digital LUT-based calibration or transistor intrinsic matching.

- A two-step open-loop VCO-based ADC architecture has been proposed and implemented in Chap. 6. Only two frequency-type VCO-based quantizers and one current-steering DAC are needed for the ADC design. The VCO nonlinearities have been mitigated by distortion cancellation and input-swing reduction techniques. The DAC matching requirement is released by the implicit DEM of the quantizer. Because of the compact architecture and the highly-digital building blocks with relaxed performance requirements, the proposed ADC prototype in 40 nm CMOS reaches an excellent FoM of 42fJ/Step.
- Chapter 7 has investigated the VCO-based MASH $\Delta\Sigma$ ADC architecture. A power-efficient nonlinearity-cancellation technique in a 0-$\Delta\Sigma$ MASH structure has been analyzed. This chapter has covered the full design procedure of a 0–2 MASH VCO-based ADC prototype. The measurement results in 40 nm CMOS achieves a FoM of 35fJ/Step, which is the best among high-bandwidth(BW$\geq$20 MHz) $\Delta\Sigma$ ADCs.
- Conceptually, a VCO accumulates the excess phase like an ideal integrator. It is therefore beneficial to shift the analog integrator to the time domain as well. The principle of a fully-VCO-based integrator has been discussed in Chap. 8. Based on this technique, a 0–2 MASH CT $\Delta\Sigma$ ADC with fully-VCO-based integrator and quantizers has been presented, with circuit design details and silicon implementation results. This high-order noise shaping ADC is entirely digital and hence can benefit from CMOS technology scaling.

Based on this research work, several conclusions can be draw for the design of power-efficient highly-digital ADCs.

- In future communications, the RF and analog parts will be simpler and the digital core will become more important and more powerful. As a result, ADCs will be closer to the receiver antenna and play a more critical role in the overall performance of the whole system.
- With technology scaling, analog circuits suffering from the reduced voltage dynamic range cannot fully benefit from this scaling. So digitization is one important design trend for ADCs, for both speed and power considerations. There will be less analog blocks, more digital blocks (e.g. digital calibration) and semi-digital blocks (e.g. comparator, DAC and VCO) used in future ADC design.
- Using an advanced technology is always beneficial for high-bandwidth low-power DSM ADC designs, especially for highly-digital architectures. At the same time several problems become more severe when using these technologies, including a low supply voltage, a low amplifier gain and bulk/gate leakage currents. To resolve such challenges, innovations in both ADC architectures and circuit blocks are required, as presented in this book.
- VCO-based ADC is one of the most promising candidates due to its simple structure and inherent 1st-order noise-shaping characteristic. However, the performance of VCO-based ADCs is severely limited by the inherent nonlinearity of the VCOs, which needs to be solved by design innovations.

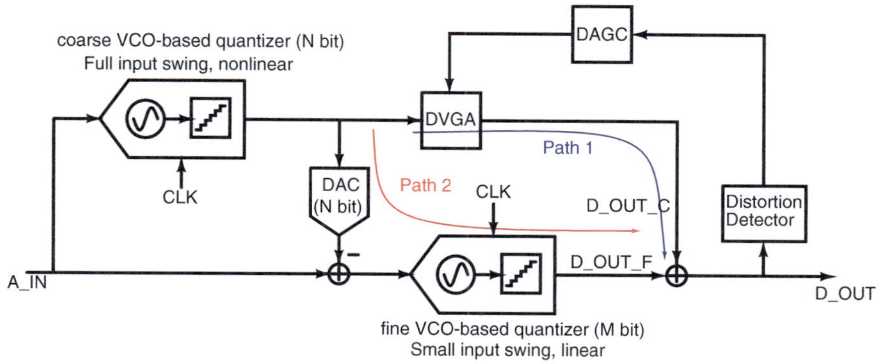

Fig. 9.1 Block diagram of the digital calibration for the two-step open-loop VCO-based ADC

- A power-efficient two-step open-loop VCO-based ADC architecture has been proposed in this book to mitigate the VCO-based ADC nonlinearity.
- A power-efficient nonlinearity-cancellation technique in a 0-$\Delta\Sigma$ MASH ADC structure has been proposed in this book to solve the performance bottleneck of the VCO-based ADCs.
- The power-hungry analog integrators can also be replaced by VCOs. Therefore, high-order CT $\Delta\Sigma$ ADCs can be implemented with fully-VCO-based integrators and quantizers, without any analog amplifier.
- In this work, low-power (FoM as low as 35fJ/Step) highly-digital $\Delta\Sigma$ ADCs with moderate accuracy (10–12 bit) have been demonstrated in 90 and 40 nm CMOS technologies. They are suitable for next-generation communication systems with performance-relaxed RF and analog front-ends.

9.2 Suggestions for Future Work

As presented in this book, research has been carried out on power-efficient highly-digitally $\Delta\Sigma$ ADC designs, using tens of MHz bandwidth as example in this book. However, this is only one brick of the whole wall. Several research directions can be explored further in the future to improve the ADC specifications and to develop the design closer to a real product.

- **Digital calibration for the two-step open-loop VCO-based ADC**. In the two-step open-loop VCO-based ADC presented in Chap. 6, the linearity performance of the whole ADC depends on the matching of two paths. Now the gain of one path is tuned manually, by adjusting the VCO gain (coarse tuning) and the DAC bias current (fine tuning). Since the outputs of both paths are digital, digital calibration can easily be included to solve the path matching automatically, as shown in Fig. 9.1. First, a digital harmonic detector is used to sense the ADC output distortion. Then a simple decision unit is assigned to determine whether to increase or decrease the gain of the DAC fine quantizer path. For the gain tuning, two methods

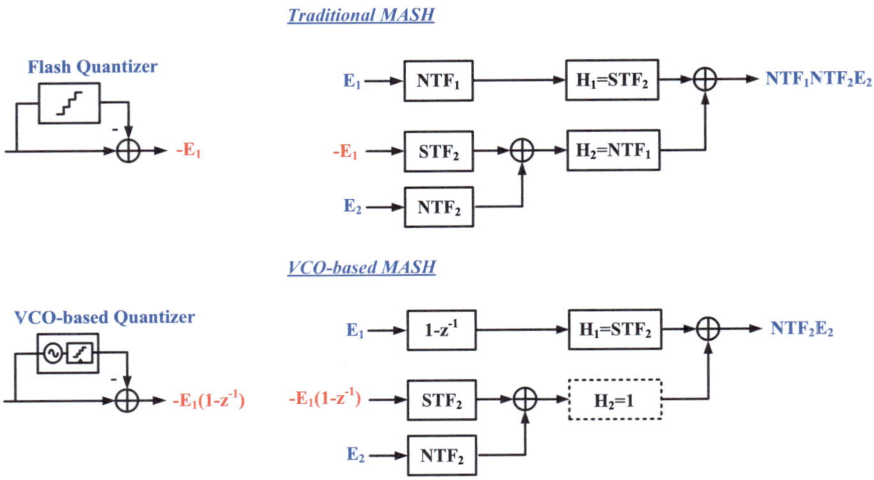

Fig. 9.2 Topology comparison between the traditional MASH ADC and the presented VCO-based MASH ADC

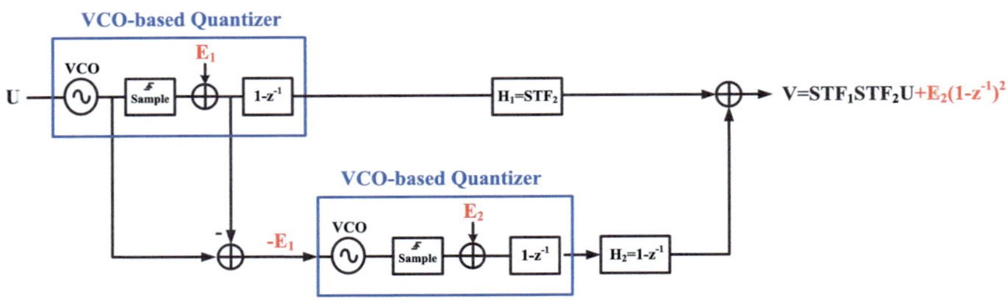

Fig. 9.3 Improved 1–1 MASH structure for VCO-based ADC

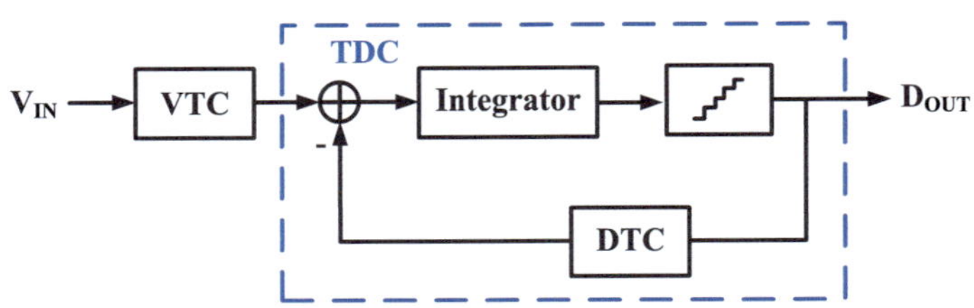

Fig. 9.4 Block diagram of a purely-time-domain $\Delta\Sigma$ ADC

can be applied. One option is to add some digital control bits to the DAC and the fine VCO, which are controlled by the decision unit outputs. Another choice is to insert a digital reconfigurable-gain stage between the coarse quantizer and the adder, and the gain is set by the decision unit. This digital calibration scheme is similar to the noise cancellation filter used in CT cascade DSM ADCs [1] and the complex image rejection circuit used in a low-IF receiver [2].

- **Bandpass/Quadrature DSM ADC for low-IF, IF and RF signal A/D conversions**. Today most DSMs are designed for baseband signal A/D conversions. With the increasing requirement for system programmability and the advances in receiver architectures, more and more A/D conversions are done in the IF or even the RF in the receiver signal chain. Among all ADC topologies, the DSM is the only architecture which can do bandpass signal A/D conversion, making it a perfect choice for IF and RF signal conversions. If the central frequency is not so high, the basic resonator can still be implemented by local feedback across two active-RC or gm-C integrators. If the IF/RF signal frequency is much higher, then an active-LC resonator is needed in the design. In such a design, high-quality on-chip inductors is a design challenge. VCO-based bandpass ADC for RF sampling has also been reported [3] without feedback loop, showing design merit of time-domain quantization in bandpass ADCs. One important difference between the bandpass and the baseband A/D converters is that the bandpass ADCs are much more sensitive to clock jitter. A lot of research is still needed to achieve bandpass A/D conversion with a high central frequency and a high conversion accuracy.
- **Improved MASH structure for VCO-based ADC**. Considering the ADC designs presented from Chaps. 6 to 8, although the first VCO-based stage is 1st-order noise shaped, it cannot provide additional noise shaping for the whole ADC, therefore these prototypes are regarded as 0–1 or 0–2 MASH ADCs. The reason is summarized in Fig. 9.2. For the presented VCO-based MASH ADC, the input of the second stage is $E_1(1 - z^{-1})$ instead of E_1 for the traditional MASH structure. To cancel out E_1 at the final output, the digital filter H_2 should be set to 1 instead of NTF_1. As a result, for the quantization noise E_2 of the second stage, only NTF_2 provides a shaping capability. To improve the ADC performance, an improved VCO-based MASH structure as shown in Fig. 9.3 can be implemented. If the phase error of the first VCO-based stage can be extracted, the new structure is similar to a conventional 1–1 MASH ADC, where the digital filter H_2 is set as NTF_1. The final output V can be expressed as:

$$V = STF_1 \cdot STF_2 \cdot U + NTF_1 \cdot NTF_2 \cdot E_2 \tag{9.1}$$

  As observed from the above analysis, an important advantage of this topology is that a higher-order noise shaping can be achieved by cascading open-loop VCO-based stages. Therefore, the power-hungry analog integrators can be removed, which is beneficial for high-speed and low-power applications. Reference [4] provides a potential design example of an improved 2nd-order 1–1 MASH VCO-based ADC, in which the front-end PWM pre-coding can be removed.
- **Purely-time-domain $\Delta\Sigma$ ADCs**. Reviewing the fully-VCO-based ADC design presented in Chap. 8, the VCO-based integrator transfers the voltage-based signal into the phase domain, and then recovers it back to the voltage domain by a current-steering DAC. Therefore, the signal is switched from the voltage domain to the time domain back and forth, causing extra power to be dissipated in the transfer. If the signal can be processed purely in the time domain, the whole ADC will be more digital and power-efficient. The structure is illustrated in Fig. 9.4. After the

voltage-to-time conversion, a $\Delta\Sigma$ closed loop is implemented in the time domain. References [5, 6] provide two potential design examples, both of which utilize a VCO as the voltage-to-time converter (VTC). To realize the $\Delta\Sigma$ function entirely in the time domain, [5] implements a phase interpolation, while [6] adopts a up-down counter. In this kind of structure, the ADC performance is limited by the nonidealities of the front-end VTC.

This research work has achieved and demonstrated power-efficient highly-digital DSM A/D conversions in nanoscale CMOS technologies.

References

1. L.J. Breems, R. Rutten, G. Wetzker, A cascaded continuous-time $\Delta\Sigma$ modulator with 67-dB dynamic range in 10-MHz bandwidth. IEEE J. Solid State Circuits **39**(12), 2152–2160 (2004)
2. S. Lerstaveesin, Bang-Sup song, a complex image rejection circuit with sign detection only. IEEE J. Solid State Circuits **41**(12), 2693–2702 (2006)
3. Y.G. Yoon, S. Cho, A 1.5-GHz 63dB SNR 20mW direct RF sampling bandpass VCO-based ADC in 65nm CMOS, in *Symposium on VLSI Circuits*, vol. 2009 (Kyoto, Japan, 2009), pp. 270–271
4. P. Gao, X. Xing, J. Craninckx, G. Gielen, Design of an intrinsically-linear double-VCO-based ADC with 2nd-order noise shaping, in design. Autom. Test Eur. Confer. Exhib. (DATE) **2012**, 1215–1220 (2012)
5. M. Amin, B. Leung, Design techniques for linearity in time-based SD analog-to-digital converter. IEEE Trans. Circuits Syst. II Expr. Briefs **63**(5), 433–437 (2016)
6. A. Babaie-Fishani, P. Rombouts, True high-order VCO-based ADC. Electr. Lett. **51**(1), 23–25 (2015)

Index

X. Xing et al., *Design of Power-Efficient Highly Digital Analog-to-Digital Converters for Next-Generation Wireless Communication Systems*, Signals and Communication Technology, https://doi.org/10.1007/978-3-319-66565-8

MIX
Papier aus verantwortungsvollen Quellen
Paper from responsible sources
FSC® C105338

If you have any concerns about our products,
you can contact us on
ProductSafety@springernature.com

In case Publisher is established outside the EU,
the EU authorized representative is:
Springer Nature Customer Service Center GmbH
Europaplatz 3, 69115 Heidelberg, Germany

Printed by Libri Plureos GmbH
in Hamburg, Germany